Genes, Cells

A·N·D

Organisms

GREAT BOOKS IN EXPERIMENTAL BIOLOGY

EDITED BY
John A. Moore

A Garland Series

·

17 TITLES THAT STAND AS
MONUMENTS OF BIOLOGICAL THOUGHT

GREAT BOOKS IN EXPERIMENTAL BIOLOGY

1. *THE CELL IN DEVELOPMENT AND HEREDITY*,
   by Edmund B. Wilson.
   1925.

2. *THE CELL THEORY: A RESTATEMENT, HISTORY, AND CRITIQUE*,
   by John R. Baker.
   1948–1955.

3. *SPECIES AND VARIETIES, THEIR ORIGIN BY MUTATION: LECTURES DELIVERED AT THE UNIVERSITY OF CALIFORNIA*,
   by Hugo De Vries.
   3rd edition, 1912.

4. *THE GENETICS OF DROSOPHILA*,
   by Thomas Hunt Morgan (with Calvin B. Bridges and A. H. Sturtevant).
   1925.

5. *THE THEORY OF THE GENE*,
   by Thomas Hunt Morgan.
   1926.

6. *AN INTRODUCTION TO GENETICS*,
   by A. H. Sturtevant and G. W. Beadle.
   1939.

7. *RECENT ADVANCES IN CYTOLOGY*,
   by C. D. Darlington.
   1937.

8. *PHYSIOLOGICAL GENETICS*,
   by Richard Goldschmidt.
   1938

9. *INVERTEBRATE EMBRYOLOGY*,
   edited by Matazo Kume and Katsuma Dan.
   1968.

10. *EMBRYONIC DEVELOPMENT AND INDUCTION*,
    by Hans Spemann.
    1938.

11. *THE ORGANIZATION OF THE DEVELOPING EMBRYO*,
    by Ross G. Harrison.
    1969.

12. *PROBLEMS OF FERTILIZATION*,
    by Frank R. Lillie.
    1919.

13. *THE BIOLOGY OF THE CELL SURFACE*,
    by Ernest Just.
    1939.

14. *FEATURES IN THE ARCHITECTURE OF PHYSIOLOGICAL FUNCTION*,
    by Joseph Barcroft.
    1934.

15. *THE NEUROPHYSIOLOGICAL BASIS OF MIND; THE PRINCIPLES OF NEUROPHYSIOLOGY*,
    by John Carew Eccles.
    1953.

16. *RECOLLECTIONS OF MY LIFE*,
    by Santiago Ramon Y. Cajal.
    1937.

17. *THE DISCOVERY AND CHARACTERIZATION OF TRANSPOSABLE ELEMENTS: THE COLLECTED PAPERS OF BARBARA McCLINTOCK*.
    1987.

THE CELL THEORY

A Restatement, History, and Critique

John R. Baker

Garland Publishing, Inc.
NEW YORK & LONDON • 1988

Library of Congress Cataloging-in-Publication Data

Baker, John Randal, 1900–
The cell theory : a restatement, history, and critique / John R. Baker.
p. cm.—(Genes, cells, and organisms)
Collection of articles originally published in Quarterly journal of microscopical science, 1948–1955.
ISBN 0-8240-1388-3
1. Cytology. 2. Cytology—History. I. Title. II. Series.
QH581.2.B35 1988 88-23428
547.87—dc19

All volumes in this series are printed on acid-free, 250-year-life paper.

PRINTED IN THE UNITED STATES OF AMERICA

THE CELL THEORY:
A RESTATEMENT, HISTORY, AND CRITIQUE
has been collected and reprinted from five issues of the *Quarterly Journal of Microscopical Science*, 1948-1955.

Part I:	*Q.J.M.S.,*	Vol. 80,	Third Series, No. 1	103-125
Part II:	*Q.J.M.S.,*	Vol. 90,	Part 1, March 1949	87-108
Addendum to Part II		90,	Part 3	331
Part III:	*Q.J.M.S.,*	Vol. 93,	Part 2, June 1952	157-190
Part IV:	*Q.J.M.S.,*	Vol. 94,	Part 4, Dec. 1953	407-440
Part V:	*Q.J.M.S.,*	Vol. 96,	Part 4, Dec. 1955	449-481

The Cell-theory: a Restatement, History, and Critique

PART I

BY

JOHN R. BAKER

(From the Department of Zoology and Comparative Anatomy, Oxford)

CONTENTS

INTRODUCTION

SEVERAL zoological text-books published during the last two decades have cast doubts on the validity of the cell-theory. These books do not present in any comprehensible form the evidence on which the doubts are based. I therefore set myself the task of finding and studying the evidence, in order to be able to form some judgement of its weight. When the evidence was examined, it became apparent that different parts of the theory were being attacked, and that one attack might be justifiable while another was not.

Now there is no generally accepted body of opinion as to what the cell-theory is. The phrase 'cell-theory' was invented by Schwann, who has told us what he meant by it (1839*a*, p. 197). 'One may include under the name of *cell-theory*, in the wider sense,' he wrote, 'the exposition of the statement that there exists a general principle of construction (Bildungsprinzip) for all organic products, and that this principle of construction is cell-formation.' Unfortunately, the word Bildungs introduces uncertainty into the meaning of Schwann's definition, for it may refer either to *structure* or to *development of structure*. He seems to have meant the latter. Schwann thought that the development of structure took place in two stages: the development of cells from a structureless substance, and the differentiation of those cells into their definitive forms. He expressed this best in a little-known passage (Schwann, 1839*b*, p. 139), contributed to a work by another author. He here gives a clearer and better definition of the cell-theory than in his celebrated book.

'A common principle of development (Entwickelungsprinzip)', he wrote, 'is the basis of all organic tissues, however diverse they may be, namely, cell-formation (Zellenbildung); that is to say, nature never joins the molecules together directly into a fibre, tube, &c., but always first fashions a cell or first transforms this cell, where necessary, into the different elements of structure as they occur in the adult state.'

Schwann's opinion as to the first stage of the development of structure was quite wrong. In October 1837 he took over from Schleiden, in conversation (Schwann, 1839*a*, p. x; 1884, p. 25), an entirely false theory as to how cells are formed. Following Schleiden, he called the nucleus a Cytoblast or cell-bud. He thought that nuclei were formed by a process resembling crystallization in a structureless fluid, the Cytoblastem (or cytoblastema, as it is often spelled in English); that each secreted a membrane round itself; and that what we nowadays call the cytoplasm then appeared within this membrane. The method by which cells are formed, however, was not a subject that Schwann investigated with any thoroughness: his research was devoted to the structure of cells, rather than to the process by which they originate. For instance, the cartilage-cell was one of his first objects of study in connexion with the cell-theory (Schwann, 1838*a*), and he laid especial stress on it in his book (1839*a*); yet his evidence for the view that it develops from a Cytoblastem in accordance with Schleiden's scheme is wholly indirect. This is made particularly clear in the English edition of his book (1847), for which he rewrote part of the section dealing with cartilage. Nevertheless, Schwann not only believed that animal cells resemble those of plants in their mode of development from a structureless Cytoblastem, but thought that it was he who had discovered this, and insisted on his priority in the matter in the course of an argument with Valentin, attached to the end of his book in both editions.

Schwann drew a sharp distinction between his cell-theory (Zellentheorie) and his theory of the cells (Theorie der Zellen). The former he regarded as inductive, the latter as speculative. In his theory of the cells he is concerned with teleology, with what we should nowadays call the colloid chemistry of cells, and with cellular in relation to organismal individuality. Schwann himself did not regard the last-named subject as part of his cell-theory.

Remak (1855, p. 164), who attributed the cell-theory to Schleiden, defined it as 'the theory of the formation of plants exclusively from homologous components which develop in different ways'. He thus laid stress on the conformity of all plant cells with one another, rather than on the particular way in which they develop. Virchow (1859, p. 9), who was well aware that cells originate from pre-existing cells and who coined his famous aphorism to call attention to this fact, took the phrase 'cell-theory' to mean precisely the erroneous views of Schleiden and Schwann as to their origin.

Other authorities have laid particular stress on the double individuality that they believe to be characteristic of many-celled organisms. Schleiden himself (1838, p. 138) wrote: 'Now each cell leads a double life: the one wholly independent, only connected with its own development, and the other remote, in so far as it has become an integral part of a plant.' Hertwig (1893, p. 3), in defining the cell-theory, added this idea of individuality to a clause that is reminiscent of Schwann and Remak: 'Animals and plants, so diverse in their external appearance, agree in the fundamental nature of their anatomical construction; for both are composed of similar *elementary units*, which are generally only perceptible under the microscope. Through the influence of an old theory, now discarded, these units are called cells, and thus the doctrine that animals and plants are composed in an accordant manner of very small particles of this kind is called the *cell-theory* . . . the common life-process of a composite organism appears to be nothing else than the exceedingly complicated result of its numerous and diversely-functioning cells.' Bourne (1895, p. 162) also seems to be influenced by Schleiden's ideas when he sums up the cell-theory thus: 'The multicellular organism is a colony, consisting of an aggregation of separate elementary parts, viz. cells. The cells are independent life units, and the organism subsists in its parts and in the harmonious interaction of those parts.' (Bourne himself only accepted a part of this theory as true.) In recent times, Karling (1939, p. 525) has expressed himself similarly: 'The concept of the organism as a mass of cells which integrate and interact to form a co-ordinate whole is perhaps the real climax of the theory.' Other authors have regarded the homology of the cells of many-celled organisms with individual protists as an essential part of the cell-theory. This opinion as to the meaning of the theory has been held by some of those who attack it (e.g. Dobell, 1911). Indeed, the attack on the cell-theory has been mainly directed against this aspect of it.

The diversity of views about the meaning of the expression 'cell-theory' makes it evident that the truth or untruth of the theory can only be established if it can be formulated in clear terms. The different parts of the cell-theory do not necessarily stand or fall together, and the formulation should therefore be in a series of separate propositions, each of which can be examined independently. No proposition which was formerly thought to be a part of the cell-theory, but which is universally discarded by modern biologists, need be included in the formulation; for instance, it would be useless to set up Schleiden's view as to the mode of origin of cells and then demolish it.

Bearing these considerations in mind, I restate the cell-theory in a series of seven propositions, as follows:

I. Most organisms contain or consist of a large number of microscopical bodies called 'cells', which, in the less differentiated tissues, tend to be polyhedral or nearly spherical.

II. Cells have certain definable characters. These characters show that cells (*a*) are all of essentially the same nature and (*b*) are *units* of structure.

III. Cells always arise, directly or indirectly, from pre-existing cells, usually by binary fission.

IV. Cells sometimes become transformed into bodies no longer possessing all the characters of cells. Cells (together with these transformed cells, if present) are the living parts of organisms: that is, the parts to which the synthesis of new material is due. Cellular organisms consist of nothing except cells, transformed cells, and material extruded by cells and by transformed cells (except that in some cases water, with its dissolved substances, is taken directly from the environment into the coelom or other intercellular spaces).

V. Cells are to some extent individuals, and there are therefore two grades of individuality in most organisms: that of the cells, and that of the organism as a whole.

VI. Each cell of a many-celled organism corresponds in certain respects to the whole body of a simple protist.

VII. Many-celled plants and animals probably originated by the adherence of protist individuals after division.

In considering the evidence for and against each proposition, it was necessary to discover how the various bodies of opinion originated and developed. The history of the cell-theory has been told many times, not only in the indispensable standard text-books of the history of biology, but also in other books and in papers. The historical studies that have been of particular use to me are those of Burnett (1853), Tyson (1870 and 1878), M'Kendrick (1888), Turner (1890 *a* and *b*), Sachs (1890), Hertwig (1893), Karling (1939), and Wilson (1944); those of Mark (1879–80), Strasburger (1880), Waldeyer (1888), Rádl (1930), Conklin (1939), and Woodruff (1939) have also been helpful. Among all the existing literature, however, I have not found a sufficiently detailed or accurate study of the origin and development of the various bodies of opinion. What has been written previously on the subject has helped me mainly by leading, directly or indirectly, to original sources. I have not relied and shall not rely in a single instance on a statement by one author as to what another says. A considerable amount of research has been necessary, which has shed new light on some parts of the cell-theory and contradicted certain generally accepted opinions.

In seeking to affect opinion, scientists are usually careful to make their conclusions verifiable with the least possible trouble. It would be a pleasure to be able to say the same of historians of science, but unfortunately it would also be an unpardonable exaggeration. I am creating something of a precedent by giving exact page-references to most of the sources of my information, except when they are contained in such short papers that this is unnecessary. I intend in this way to make it as easy as possible for readers to check the accuracy of what I say and to correct any errors. All verbal quotations given in this series of papers will be rendered in English. The exact references will make it easy to find the originals in every case. In translating from the various languages, I have tried to be as literal as is consistent with the writing of genuine English: no attempt has been made to preserve any foreign grammatical forms and thus produce the kind of half-translation that is familiar in scientific literature. I have not relied on the translations of others, except where the author's manuscript was translated before publication, or where the original was not available to me for some other reason. In the case of Swammerdam's *Biblia naturae* (1737–8), which was written by the author in Dutch but printed in Latin and Dutch in parallel columns, I have used the Latin version (except where the contrary is stated). When a printed original source has not been available, I have made it clear that the translation used was not my own, and have given a reference to the translation instead of to the original.

The books used in this investigation have been provided by the Radcliffe and Bodleian libraries and those of the Department of Botany at Oxford, of the Royal Society, the Royal Geographical Society, the British Museum, and the British Museum (Natural History).

I hope that readers who may disagree with the conclusions I draw as to the validity or invalidity of the several propositions may, nevertheless, find some value in the historical parts of this series of papers.

The attempt will be made to express all ideas with the utmost clarity and simplicity, so that there will be no mistake about my meaning. Then, if the ideas are wrong, they can easily be corrected.

In composing the papers I have received help from the criticisms of the cell-theory by Whitman (1893), Sedgwick (1894 and 1895), Bourne (1895, 1896 *a*, *b*, and *c*), Awerinzew (1910), Dobell (1911), Baitsell (1940), Weiss (1940), and Ries (1943). I acknowledge the assistance given by friends in the course of informal discussions. In particular I must mention the encouragement given by Prof. A. C. Hardy, F.R.S., which has helped me in a long and rather difficult task, somewhat removed from the main stream of modern cytological advance.

Proposition I

Most organisms contain or consist of a large number of microscopical bodies called 'cells', which, in the less differentiated tissues, tend to be polyhedral or nearly spherical.

The Discovery of Plant Cells

Strangely enough, the earliest published account of the microscopical structure of plant tissues is concerned with petrified wood. In describing the resemblances of this material to ordinary wood, Hooke refers to the 'conspicuous *pores*'. He continues: 'Next (it resembled wood) in that all the smaller and (if so I may call those which are only to be seen by a good glass) *microscopical* pores of it, appear (both when the substance is cut and polish'd *transversly*, and *parallel* to the pores) perfectly like the *Microscopical* pores of several kinds of *wood*, retaining both the shape, and position of such pores.' The 'conspicuous *pores*' may have been either resin-canals or large vessels. Most of the '*microscopical* pores' were presumably small vessels, though some of them may have been cells of the medullary rays or of the wood parenchyma. Hooke regarded them as tubular. He contributed his account of the structure of petrified wood to Evelyn's *Sylva* (1664, p. 96).

Evelyn spells Hooke's name 'Hook', but there is no doubt about the identity of the famous microscopist, for Hooke refers to his study of petrified wood again in his *Micrographia* (1665, p. 100). In this work, before describing cork, Hooke writes '*Of* Charcoal, *or burnt* Vegetables.' In describing the stems of plants, burnt to charcoal and broken across, he first notices the large vessels and proceeds (p. 101): 'But this is not all, for besides those many great and conspicuous irregular spots or pores, if a better *Microscope* be made use of, there will appear an infinite company of exceedingly small, and very regular pores, so thick and so orderly set, and so close to one another, that they leave very little room or space between them to be fill'd with a solid body, for the apparent *interstitia*, or separating sides of these pores seem so thin in some places, that the texture of a Honey-comb cannot be more porous. Though this be not every where so, the intercurrent partitions in some places being very much thicker in proportion to the holes.

'Most of these small pores seem'd to be pretty round, and were rang'd in rows that radiated from the pith to the bark.' He considered the 'pores' to be longitudinal tubes. He calculated that there are 2,700 of them to an inch (in a transverse section). He mentions that in sound wood the microscopical pores 'are fill'd with the natural or innate juices of those Vegetables' (p. 108).

After dealing with charcoal and repeating his already-published observations on petrified wood, Hooke comes to his well-known study '*Of the* Schematisme *or* Texture *of* Cork'. He appears to have seen cells in cork before he studied petrified wood and charcoal, for he writes (pp. 112–15) that the cells of cork 'were indeed the first *microscopical* pores I ever saw, and perhaps, that were ever seen, for I had not met with any Writer or Person, that had made any mention of them before this'. He describes how he cut a very thin section of cork and examined it on a black plate with a plano-convex lens. '. . . I could exceeding plainly perceive it to be all perforated and porous, much like a Honey-comb, but that the pores of it were not regular; yet it was not unlike a Honey-comb in these particulars . . . these pores, or cells, were not very deep, but consisted of a great many little Boxes. . . . Nor is this kind of Texture peculiar to Cork onely; for upon examination with my *Microscope*, I have found that the pith of an Elder, or almost any other Tree, the inner pulp or pith of the Cany hollow stalks of several other Vegetables: as of Fennel, Carrets, Daucus, Bur-docks, Teasels, Fearn, some kinds of Reeds, &c. have much such a kind of *Schematisme*, as I have lately shown [in] that of Cork.' Hooke thought that the 'pith' of the shaft of a feather had a similar structure.

Unlike some of his followers, Hooke did not concentrate the whole of his attention upon the cell-wall. He wrote (p. 116): '. . . in several of those Vegetables, whilst green, I have with my *Microscope*, plainly enough discover'd these Cells or Poles [misprint for Pores] filled with juices . . . as I have also observed in green Wood all those long *Microscopical* pores which appear in Charcoal perfectly empty of any thing but Air.' This passage provides the earliest mention of the substance of cells (as apart from that of cell-walls), though Hooke naturally did not distinguish between cell-substance and sap.

Grew made a study of the microscopical structure of plants independently of Hooke, and saw the cells. He had finished the 'composure' of his little book (Grew, 1672) when Hooke communicated his observations to him. Grew had supposed that the pith of plants consisted partly of cells like those of a honeycomb and partly of long tubes. Hooke now corrected him on this matter and the correction was accepted (p. 78). Even in this little book, Grew carried the study of plant cells much farther than Hooke. He showed (pp. 78–9) the cellular nature of the cortex as well as the pith and illustrated his findings by a figure (his Fig. 15) of the cortex of the stem of the burdock (*Arctium*) in section. This is probably the earliest-published figure of cells as they occur in a living plant (for Hooke's figures were of petrified wood, charcoal, and cork). Grew carried his researches into a still more important field when he demonstrated the cellular nature of plant embryos. He describes the 'sameness' of the nature of the pith and cortex 'with the *Parenchyma* of the Seed. For, upon farther enquiry with better Glasses, I find, that the *Parenchyma* of the *Plume* and *Radicle*, and even of the *Lobes* themselves, though not so apparently, is nothing else but a Mass of Bubbles' (p. 79).

Ten years later, Grew (1682) published his great book on *The Anatomy of Plants*. The figures, many of them representing microscopical dissections in three dimensions like a modern stereogram, are magnificent. In this work he uses the words 'Bladders', 'Cells', and 'Pores' indiscriminately (see, for example, p. 64). He had now acquired, however, a wrong idea of the units of plant structure. His 'Mass of Bubbles' was a much better analogy than that of '*fine Bone-Lace*' (p. 121), which he develops in a celebrated passage in his later work. This subject will be reviewed under the heading of the Proposition II in the second part of this series of papers.

Meanwhile, Malpighi had been busy with the same subject. At the end of his work, *Anatomes plantarum idea*, he wrote the date 1671. In this paper, which is not illustrated, Malpighi calls cells 'utriculi' and 'sacculi'. The *Idea* was first published four years after it was written, in the large volume, *Anatome plantarum* (Malpighi, 1675). This book included more detailed studies of microscopical plant anatomy. The cellular structure of plants is illustrated by figures of transverse and longitudinal sections of stems. These figures do not approach those of Grew (1682) in elaboration of detail. There has been some discussion as to whether Grew or Malpighi had priority in the microscopical study of plant tissues, and whether Grew was more indebted to Malpighi or vice versa. Schleiden (1849, p. 37) was in error when he said that Malpighi despatched his *Idea* to the Royal Society in 1670. He did not do so until November of the next year. Meanwhile, on 1 May 1671 the Royal Society had already given the order for the printing of Grew's little book (1672). Pollender (1868), who has carefully studied the question of priority, awards it unhesitatingly to Grew (p. 7), while allowing that Malpighi's early work was done quite independently.

Leeuwenhoek was not slow to follow up the work of these early investigators of plant cells, to whom he refers, in an undated letter to Hooke, as 'acutissimos Viros *Malpighium* & *Nehemiam Grew*' (see Leeuwenhoek, 1722, p. 13). Leeuwenhoek here gives a large drawing of part of a transverse section through a stem of oak. His writings on plant cells, however, will not bear comparison with those of Grew, and scarcely with those of Malpighi. On this subject he exhibits to the full his character of dilettante of genius. He repeatedly describes and figures the cells of plants, and continues to do so up till near the end of his life. In a letter written in 1713, for instance, he figures them in the seeds of various plants (see Leeuwenhoek, 1719, pp. 25–6). From his time onwards it was a matter of common knowledge among botanists that plants were constructed largely of microscopical chambers, though vessels were not known to be of cellular origin until much later. (The cellular origin of vessels will be discussed under the heading of Proposition IV, in a later paper in this series.) The following statement by Moldenhawer (1812, p. 86) is representative of the best opinion of his time: '. . . the cellular substance thus consists of single, closed, spherical, oval or more or less oblong, almost cylindrical utricles, which, owing to mutual pressure on one another, assume an angular and flattened form, either regular or more or less irregular, and resembling the cells of a honeycomb.'

The Discovery of Blood Corpuscles

The absence of thick cell-walls in most animal tissues put zoologists at a great disadvantage, in comparison with botanists, in recognizing the cellular nature of organisms. Complete separateness of cells was the next best help to microscopists after the presence of easily visible cell-walls. The blood is the only part of an animal that can be compared with most plant tissues in the ease with which it reveals its cellular nature.

It will probably never be known with certainty by whom blood corpuscles were discovered. They were certainly seen by Swammerdam, who died in 1680. It is well known that his *Biblia naturae* was published for the first time long after his death. In this great work (Swammerdam, 1737–8, vol. 1, p. 69), he first mentions blood corpuscles in connexion with his description of the dissection of the louse (*Pediculus*): 'If we begin the dissection in the upper part of the abdomen, and cautiously split the skin there, blood immediately escapes from that place. The blood, when received into a glass tube and examined with a very good microscope, is observed to consist of transparent globules (globulis), in no way differing from cow's milk, a fact that was discovered a few years ago in human blood also; for this is seen to consist of slightly reddish globules, floating in a clear fluid.' It will be noticed that Swammerdam does not state that it was he who made the discovery in human blood. He remarks: '... I shall not recklessly assert that globules are present in the blood of the louse, for it could easily happen that fat might mix itself [with the blood], and so also might certain fragments of the viscera, damaged [by the dissection]; these certainly consist of a mass, as it were, of globular particles, as I shall show at the proper time.' He illustrates the globules, as seen within the glass tube, in Fig. 1 of Tab. II. He did not realize that they might be those of human blood sucked by the louse.

Swammerdam refers to the blood corpuscles of the frog in the second volume of the same book. He writes (p. 835): 'In the blood I saw a watery part, in which floated an immense number of circular particles, rejoicing in a flat, as it were oval, but perfectly regular shape. These particles seemed also to contain another fluid again within themselves. But if I looked at them from the side, they resembled crystalline rods and many other figures; according, doubtless, to the various ways in which they were rotated in the fluid of the blood. I observed moreover that the larger the objects were represented through the intervention of the microscope, the paler their colour appeared [to be].'

Flloyd's translation of the *Biblia* (Swammerdam, 1758) suggests strongly, on p. 120, that Malpighi and Needham already knew of the presence of globules in blood before Swammerdam. The original work was produced in Dutch and Latin in parallel columns. In the passage which Flloyd is here translating, neither the Dutch nor the Latin version gives any evidence for the suggestion that Malpighi or Needham had priority over Swammerdam in this matter.

Unfortunately there would seem to be no means of dating these observations. Foster (1901, p. 99) says that Swammerdam discovered blood corpuscles in 1658 but gives no evidence for this. Miall (1912, p. 198) called attention to the lack of evidence, and no answer appears ever to have been made. It is quite possible that Swammerdam demonstrated the blood corpuscles to the Duke of Tuscany, when the latter visited him and offered him employment. If a record of this particular demonstration exists, the attention of present-day biologists should be called to it. Swammerdam himself (1737–8, p. 839) mentions the Duke's visit, and tells how he demonstrated a nerve-muscle preparation on this occasion. Swammerdam gives 1658 as the date of the visit, but Boerhaave, in writing the great naturalist's biography as a preface to the *Biblia*, states that the date was 1668 (page facing p. C2). Stirling (1902, p. 23), like Foster, gives 1658 as the date of Swammerdam's discovery of blood corpuscles but provides no new evidence.

In his *Exercitatio de omento, pinguedine, et adiposis ductibus*, published in 1665, Malpighi makes the first mention in print of blood corpuscles (see Malpighi, 1686, vol. 2, p. 41). It cannot be claimed that the discovery was made in a satisfactory manner, for he regarded them as globules of fat; yet it is clear from his words that he saw them and that he had no previous knowledge of the existence of such objects in blood. He wrote: '... in the omentum of a hedgehog, in a blood-vessel that extended from an accumulation of fat to another opposite to it, I saw globules of fat, possessing an outline of a particular shape, and reddish; they resembled in general a circle (coronam) of red corals.'

In a letter written to the Royal Society in 1673 (or possibly in 1674), Leeuwenhoek thus describes his own discovery of blood corpuscles (Leewenhoeck [*sic*], 1674, p. 23): 'I have divers times endeavoured to see and to know, what parts the *Blood* consists of; and at length I have observ'd taking some Blood out of my own hand, that it consists of small round globuls driven through a Crystalline humidity or water: Yet, whether all Blood be such, I doubt. And exhibiting my Blood to my self in very small parcels, the globuls, yielded very little colour.' Four years later, Hooke (1678, p. 93) attributed the discovery of the 'Globules' of the blood to Leeuwenhoek.

In his characteristically random manner, Leeuwenhoek reverted to blood corpuscles from time to time. In a letter written to the President of the Royal Society in 1683 (three years after Swammerdam's death), he described and figured the red corpuscles of the frog, noting carefully how their colour appears more clearly when two or three are superimposed (see Leeuwenhoek, 1722, pp. 54–5). Writing to the Royal Society again in July 1700, he makes the first mention of the nucleus ('lumen'), and figures it, in describing the blood corpuscles of the salmon and flounder (see Leeuwenhoek, 1719, pp. 219–20). (The discovery of the nucleus will be more fully considered in the discussion of Proposition II.) Writing yet once again to the Royal Society in his old age, in 1717, Leeuwenhoek gives the first indication that human red blood corpuscles are not spheres, but concave disks (Leeuwenhoek, 1719, pp. 421–2). Sixty years later, Hewson (1777, p. 15) observed that mammalian blood corpuscles were not spherical, as was still commonly supposed, but flat, and therefore, he concluded, not fluid.

Although blood corpuscles were discovered at about the same time as the cells of plants, and both soon became familiar objects, it did not occur to the early microscopists that there was any relation between them. It was necessary first to understand that the non-fluid tissues of animals also consisted of immense numbers of minute microscopical bodies. This understanding came, in a roundabout way, through the globule-theory. But before we can follow the tortuous course of progress, it is necessary to clear away a fallacy that has misguided historians of science and given a false impression of the background of the cell-theory.

The 'tissu cellulaire' Fallacy

The expression 'tissu cellulaire', or its counterpart in other languages, occurs frequently in eighteenth- and early nineteenth-century writings on the tissues of animals. Alighting casually upon this phrase, one may easily fall into the error of supposing that it refers to cellular tissue, in the modern sense of the word 'cell'. Historians of the cell-theory, among them Gerould (1922), have been misled by this fallacy. Biologists owe a particular debt to Turner (1890*a*) and Wilson (1944) for calling attention to this matter.

Haller (1757) devotes Sectio II of his *Elementa physiologiæ corporis humani* to what he calls cellular tissue ('Tela cellulosa'). One has only to read what he writes on this subject (p. 9) to realize that he is using the phrase to mean what we should call areolar connective tissue. This appears even more evidently from his 'First lines of physiology' (Haller, 1779). The translator from the Latin of the 1766 edition renders the relevant passage as follows (pp. 3–8):

'The second kind of fibres . . . when loosely interwoven with each other, are called the *cellular* tunic; though the name *tunic* or *membrane* is on many accounts very improper.

'This cellular substance is made up of an infinite number of little plates or scales, which, by their various directions, intercept small cells and web-like spaces; and join together all parts of the human body in such a manner, as not only sustains, but allows them a free and ample motion at the same time. But in this web-like substance there is the greatest diversity, in respect of the proportion betwixt the solid parts and intercepted cells, as well as the breadth and strength of the little plates, and the nature of the contained liquor, which is sometimes more watery, and sometimes more oily. . . . This cellular web-like substance in the human body is found throughout the whole, namely, wherever any vessel or moving muscular fibre can be traced; and this without the least exception that I know of. . . . The principal use of the cellular fabric is to bind together the contiguous membranes, vessels, and fibres, in such a manner as to allow them a due or limited motion. . . . The intervals or spaces betwixt the plates or scales of the cellular membrane, are every where open, and agree in forming one continuous cavity throughout the whole body.'

It is clear that the 'cells' referred to in this passage have nothing to do with cells in the modern sense: they are simply the areolae of connective tissue.

Gerould (1922), who attributes the cell-theory to Lamarck (1809), has been misled by the words 'tissu cellulaire'. There are, indeed, some passages in Lamarck's *Philosophie zoologique* from which one might conclude that he was referring to cellular tissue in the modern sense. For instance, he says 'that the whole operation of nature for the formation of her direct creations, consists in organizing "*en tissu cellulaire*" the little masses of gelatinous or mucilaginous matter that she finds at her disposal and favourable in the circumstances' (1809, vol. 1, p. 373). A careful study of his work leaves no doubt, however, that when he refers to 'tissu cellulaire' in animal tissues, he is never referring to cells in the modern sense, but is nearly always referring to connective tissue. There are a few exceptions. In the case of polyps (vol. 1, p. 203), he appears to use the expression to mean mesogloea, while when he says (vol. 1, p. 210) that infusoria are composed of 'tissu cellulaire', we cannot guess his precise meaning. His words on p. 46 of vol. 2, however, leave no doubt whatever of the usual sense. They must be quoted in full as they settle the matter conclusively:

'. . . all the organs in animals without exception are enveloped in *tissu cellulaire*, and their lesser parts are in the same case.

'In fact, it has been recognized for a long time that the membranes that form the envelopes of the brain, of the nerves, of the vessels of all kinds, of the glands, of the viscera, of the muscles and their fibres, even the skin of the body are generally productions of *tissu cellulaire.*'

Another passage clearly indicating that 'tissu cellulaire' means connective tissue occurs on p. xiv of vol. 1. His repeated insistence (vol. 1, pp. 273 and 409; vol. 2, pp. 47 and 60) that 'tissu cellulaire' is the 'gangue' in which structure is laid down is another pointer in the same direction; for 'gangue' is the substance that encloses a metallic ore in its meshes, not the ore itself. Lamarck's ideas on morphogenesis are so unfamiliar to-day that it is difficult at first to grasp them. He makes himself clearest on this subject on pp. 373–4 of vol. 1. He regarded connective tissue as playing a fundamental part in the development of structure. Fluids move through the meshes of this tissue, and the effect of this movement is 'to open up (frayer) routes . . . to create in it canals, and consequently various organs; to vary these canals and their organs by reason of the diversity either of the movements or of the nature of the fluids'.

When dealing with plants, Lamarck uses the expression 'tissu cellulaire' to mean the cell-walls, or sometimes, more loosely, to mean the cell-walls and their contained fluids. He is therefore thinking of cells in something approaching the modern sense. It is important to realize that he homologized the cell-walls of plants with the connective tissue fibres of animals. This supposed homology seems to us so extraordinary that we do not readily understand his meaning.

Bichat (1812) devotes no less than 104 pages to the 'Système cellulaire'. What he writes at the outset (p. 11) makes it perfectly clear that by this expression he means areolar connective tissue, with the cells of which, in our modern sense, he is not at all concerned. So persistent was the term 'Zellgewebe', that Schwann himself uses it (1838*b*, col. 227) in the old sense when he wants to refer to areolar connective tissue, though in the same breath he mentions the true nucleated cells contained in it. The term 'connective tissue' is so familiar to ourselves that we may perhaps omit to reflect that it required to be invented and only gradually displaced a misleading but very familiar expression. Possibly the first use of it, in the form of 'Bindegewebe', occurs on p. 444 of Müller's *Handbuch der Physiologie* (1834), where he is discussing the histology of the kidneys, liver, &c.

The Globule-theory

What may be called the globule-theory was to some extent the forerunner of the cell-theory. Here again historians have been misled. Casual reading has suggested that various early authors knew much more about cells than in fact they did. Yet some of the 'globules' were actually cells, and to that extent the globulists were on the path of progress. The historian's difficulty is to disentangle the occasions on which they saw cells from those on which they did not.

It has already been mentioned that Swammerdam (1737–8, p. 70), in his study of the louse, stated that its viscera 'certainly consist of a mass, as it were, of globular particles (partium globulosarum), as I shall show at the proper time'. He says (p. 76) that the coats of the stomach, especially the external one, consist 'of very numerous . . . globular granules (granulis globosis)', which he describes as 'somewhat irregular'. He cannot decide whether these granules are part of the texture of the stomach or fat-particles. He remarks also (p. 70) that the muscles of the louse, when dried on glass and washed with spirits of wine, appear distinctly to be composed of globules, and also (p. 81) that the membrane that covers its nerve-ganglia seems to be composed of irregular globular particles (globosis particulis). He gives rather a confused description (pp. 84–5) of the structure of the skin, remarking that the smallest change of focus produced a new appearance. He sometimes saw 'globosae particulae' in it; sometimes it appeared to be composed of regular squares, which are illustrated in his Fig. x of Tab. II.

It is not possible to decide which of these various globules, if any, were cells and which were not. It would seem probable that he saw cells in the coats of the stomach, while the globules in the muscles may perhaps have been nuclei. Unfortunately these observations cannot be exactly dated. Boerhaave tells us in his Preface to the *Biblia* (on the page opposite p. F.2) that Swammerdam did no more scientific work after he had finished his history of the mayfly in 1675; it will be recollected that he died in 1680.

Miall (1911, p. 103), referring to the *Biblia*, says that Swammerdam 'describes a stage in which the body [of the tadpole] is entirely composed of rounded "lumps" or "granules", the *cells* of modern biology'. He repeats this on p. 106. Miall is here mistaken. On page 817 of the *Biblia* (vol. 2), Swammerdam does indeed give the impression that he knew the tadpole to be composed of cells; but it is clear from what he says at the end of the paragraph that on this occasion he is only referring to yolk-grains when he writes of 'globosas particulas'.

The year 1665 saw the first descriptions in print of what are nowadays regarded as cells of animals by Malpighi and Hooke. The former, in his *Exercitatio de omento*, refers to his microscopical examination of 'Pinguedinis globuli' and 'adiposi globuli' (reprinted in his *Opera omnia*, 1686, vol. 2, see p. 4). There would not appear to be any doubt that these were fat-cells, though it was presumably the fat-globules themselves that struck his attention, rather than the cells that contained them. Hooke (1665, p. 158) describes the hair of an Indian deer as seen under the microscope. In his figure (F in Fig. 3 of Schem. V), the imbricating scales of the cuticle of the hair are clearly seen. It looks, he says, 'like a thread of course Canvass, that has been newly unwreath'd, it being all wav'd or bended to and fro, much after that manner'. He only saw the externally projecting parts of the cuticular scales, not the complete transformed cells.

Leeuwenhoek also observed the structure of hair. He says that he examined his own hair, 'which heretofore I imagined to have seen to grow out of globuls

. . . so that Hair grows and increases by the protrusion of globuls. But two or three days agoe I observed the Hair of an *Elk*, and found it wholly to consist out of conjoyned globuls, which by my Microscope appear'd so manifestly to me, as if they could be handled' (Leewenhoeck [*sic*], 1674, pp. 23–4). Just what these globules were is uncertain; presumably they were the imbricating scales as they appeared with diffraction haloes round them.

Leeuwenhoek was now launched on his globule-finding studies, throughout which there is a curious mixture of truth and error. In the same paper he notices the 'small transparent globuls' of cow's milk, but also finds, inexplicably, that his nail consists of globules, and has no doubt that it grows from 'globuls protruded'. The most misleading of his researches in this line led to his report of the existence of globules in the brain (1686, pp. 883–9). In that of a turkey he observed 'some extream small Globules, less than 1/36th part of one of those which make the rednes in the blood'. There were also some about one-sixth of the size of a red blood corpuscle. He thought that these might have come from blood-vessels broken by himself. 'Together with the above mentioned Globules, there were some transparent irregular ones, as big or bigger than a Globule of our blood, which lay among the branches of the blood Vessels, in a space no bigger than a coarse sand.' He also found globules of various sizes in the medullary parts of the same brain, and in the brains of a sheep, ox, and sparrow.

There is nothing in this paper that could convince anyone that Leeuwenhoek saw the cells of nervous tissue. He refers more than once to his animals having been dead a considerable time before examination, and mentions (p. 885) that some of the globules seemed to consist 'of a thin transparent Oyl-like substance'. It is probable that Leeuwenhoek was looking at lipoid particles derived by maceration from cells and fibres. His description of globules in the brain, however, had a profound influence on subsequent writers.

In a letter written in 1717, Leeuwenhoek describes a transverse section of a small nerve, and gives a figure in which the fat-cells lying between the bundles of nerve-fibres are clearly seen (Leeuwenhoek, 1719, Fig. 2 on the plate opposite p. 312). He calls the fat-cells 'partes adiposae'. Monro (1726, p. 21) is probably referring to fat-cells when he describes the marrow of human bone. '. . . the Marrow,' he says, 'when hardened and viewed with a Microscope, appears like a Cluster of small Pearls. When the Oil is evaporated, these Bladders seem exactly like the *Vesiculae* of the Lungs when blown up, but not so large. The Marrow is nothing but the more oily Part of the Blood separated by the small Arteries, and deposited into these *Cellulae*.'

The foregoing observations are disconnected. With Wolff, the epigeneticist, we come to a generalization, referring to the minute structure of the embryos of animals. 'The constituent parts,' he says, 'of which all parts of the animal body are composed at their first beginnings, are globules (globuli), which always yield to a moderately good microscope' (Wolff, 1759, p. 72). He gives a figure (Fig. 1 on Tab. II) of a 28-hour chick embryo; cells can be seen in the area pellucida. (This figure is poorly reproduced in the second edition of the book (Wolff, 1774).)

Hewson (1777, pp. 63–81) investigated the microscopical structure of various glands of the lymphatic system. 'On cutting into a fresh lymphatic gland', he says, 'we find it contains a thickish, white, milky fluid. Then if we carefully wipe, or wash this fluid from any part of it, and examine it attentively in the microscope, we observe an almost infinite number of small cells, not such as have been before described, or that have been supposed to exist in the lymphatic glands, but others too small to become visible to the naked eye, expressed Plate IV. Fig. 4.' He dilutes the fluid with a solution of Glauber's salt or with blood-serum and then sees 'NUMBERLESS small, white, solid particles, resembling in size and shape those central particles found in the vesicles of the blood'. Hewson found similar 'particles' in the thymus gland and spleen. Although both his description and his figures suggest too small a size for these particles, in relation to red blood corpuscles, yet there would not seem to be much doubt that the objects of his study were lymphocytes and thymocytes. It is curious that he calls them 'cells'. He does not suggest any relation to the cells of plants.

Leeuwenhoek's ideas on the globular structure of nervous tissue had an unfortunate influence on several workers towards the end of the eighteenth and beginning of the nineteenth century. Prochaska (1779, pp. 67–8) wrote as follows: 'For when a small piece of the substance of either the cortical or the medullary part of the cerebrum or cerebellum is placed on a very thin glass and flattened out, so that it becomes conveniently transparent, then, with the help of a selected optical lens, it is revealed to be as it were a thick paste consisting of innumerable globules (globulis), in which no movement or swimming can be observed.' Prochaska tried by dilution and prolonged maceration to separate the globules completely from one another, but without success; and he concluded that they must be held together by a very subtle and very transparent connective tissue ('tela cellulari'). He describes the shape of the globules as 'not exactly spherical but irregularly rotund' (p. 70). He figures them on Figs. VIII–XI on Plate VII.

Reference has been made by a number of historians to Prochaska's work, but it does not seem to have been carefully read by them. It is certain that Prochaska did not see nerve-cells; for his globules were less than one-eighth the size of red blood corpuscles (p. 72), and they occurred in the nerves (pp. 70, 73) as well as in the brain. Despite this, it has even been said that Prochaska saw the nuclei of nerve-cells. His Fig. X does suggest this at a glance; but he describes carefully the various appearances obtained as he focused his lens on the globules, and there can be no doubt that the figure shows merely the effect of spherical aberration. This error, together with haloes produced by lenses of small numerical aperture, must often have led the globulists astray.

Fontana (1781, p. 212) also describes 'petits corpuscules ronds' in both the cortical and medullary parts of the brain. He figures them in Fig. VI of

Plate V, but it is not possible to identify them. He differed from Prochaska, however, in his account of the structure of nerves (p. 234): here he found 'cylindres tortueux primitives', which he considered to be of an elementary nature ('des principes simples primitifs, non composés d'autres moindres'). He regarded such primitive cylinders as being the basis also of tendons and muscles. He describes globules in the 'gluten' of the skin of eels. These globules, which appear to have been cells, will be described in the discussion of Proposition II. He also found globules in the retina of the eye. (Fontana repeatedly uses the phrase 'tissu cellulaire' to mean areolar connective tissue.)

Oken must be regarded as a globulist, but there is a characteristic absence of any objective descriptions that would enable his readers to check the truth of what he says. In his remarkable work, *Die Zeugung* (1805), he develops his thesis that higher organisms are constructed from a mass of infusoria. This matter will be considered further in its proper place, in the discussion of Proposition VI. A few years later (Oken, 1809, p. 26) he makes some dogmatic general remarks on the structure of organisms:

'A sphere, of which the middle is fluid but the periphery solid, is called a *bladder* (*Blase*).

'The first organic points are little bladders (Bläschen). The organic world has for its basis an infinity of little bladders.'

Many years afterwards, when the scientific world was resounding with the fame of Schwann, Oken claimed the cell-theory for himself: 'I first instituted my doctrine that all organisms arise from and consist of *little bladders* or *cells* in my book on Reproduction (Frankfurt bei Wesche, 1805, 8vo.). These little bladders, isolated and considered in their primitive origin, are the infusorial substance or primitive slime, from which all larger organisms form themselves. . . . This doctrine of the primitive constituents of the organic substance is now generally acknowledged, and I need add nothing, therefore, to the advocacy of it' (Oken, 1843, p. iii). With a characteristic stroke of genius, Oken seems to have glimpsed the homology of the cells of plants with the globules that were coming to be recognized as frequent components of animal tissues; but his writings were not very influential, because no exact observations were recorded.

Globulism reached its zenith in the brothers Josephus and Carolus Wenzel. They were clearly influenced by Prochaska, whom they quote at length. They found 'globuli', 'cellulae', or 'corpuscula' in the brains of man, rabbit, sheep, duck, fowl, pigeon, redpoll, and carp (1812, pp. 27–36). They describe them as sub-rotund, but sometimes somewhat angular. Any tendency to suppose that they actually saw nerve-cells is opposed by their finding that nerves also are composed of similar corpuscles. They found spherical corpuscles, however, in various other tissues, and some of these may have been cells. Their general conclusions are so sweeping, and seem to forestall the cell-theory in such a surprising manner, that they deserve to be quoted; but the reader should remember the weakness of the evidence on which their generalizations are based. They conclude that:

'The cortical and also the medullary substance of the human cerebrum and also of the cerebellum;

'The substance of the colliculi that are found in the interior of the human cerebrum;

'The substance of the pineal gland, of the spinal medulla, and of the nerves;

'Finally, the mass of the cerebrum in mammals, birds and fishes, consist of the same small, mutually coherent, sub-rotund corpuscles, of which the substance of muscle, liver, spleen, and kidneys is composed.'

Their final conclusion is that 'the particular structure of the whole of the cerebrum and nerves and also of all the other organs is cellular (cellulosam).... Finally, that the principle (Principium), or fundamental structure, of all the solid parts without exception is one and the same.'

Meckel (1815) was the first to incorporate the globule-theory in a text-book of anatomy. He regarded organisms as made up of two ultimate constituents: globules (Kügelchen), and 'a *coagulated*, or *coagulable* and therefore *plastic* substance' (1815, p. 4), in which the globules are invariably embedded. The globules are not always exactly spherical, but they are never angular. Apart from the blood, in which they assume particular shapes, all the globules of the body of any one species of animal have the same form; they are never elongated in one part and round in another. In man they are round. They are larger in the spleen than in the kidneys, and in the latter than in the liver. The milk globules are of the same nature as those of other organs.

Treviranus's contribution (1816) to the globular theory was not important. He remarks that all good observers see the globules (Kügelchen) in the brain. He recognizes three main elements in the tissues of animals (p. 140): elementary cylinders, protein globules (Eyweisskügelchen), and formless material. He also recognizes elementary fibres, which appear to have been connective tissue fibres, and which he regarded as of plant-like nature; but he denies (p. 126) that there is any trace in animals of anything resembling the cellular tissue of plants. It is difficult to identify Treviranus's Kügelchen; in some cases they may have been nuclei, in others artifacts, in *Hydra* nematocysts. His figures are unhelpful and show nothing that can be recognized as a cell.

If Treviranus's contribution is confused, Home's (1818, 1821) is simply erroneous. His studies were made in collaboration with a Mr. Bauer, who seems to have done most of the practical work. They noticed (1818) that when blood coagulates, the red blood corpuscles tend to unite in lines. They then boiled or roasted voluntary and involuntary muscle, macerated it in water, and found that the fibres 'are readily broken down into a mass of globules of the size of those in the blood, deprived of their colour' (1818, p. 175). They concluded that muscle is formed by the joining together of red blood corpuscles in lines, and suggested that nerve-fibres were formed in the same way. In a later paper (1821) Home describes both nerves and brain as containing innumerable globules, from 1/2,000 inch (6μ) to 1/4,000 inch in diameter. He evidently considered them to be derived from the red corpuscles of the blood. It is strange that Prevost and Dumas (1821), who were later to make an important advance in science by the discovery of cleavage, agreed with Home's opinion that muscle fibres are formed by the arrangement of red blood corpuscles in lines.

Heusinger (1822) shows affinity with Home, and indeed carries his ideas farther. He considered (pp. 113–16) that the tissues of the body had three constituents: formless matter, globules (Kugeln), and bladders (Blasen). The first he regarded 'as the mother, as the primitive sea of all other tissues'. Fibres are formed by the arrangement of globules in a row: bladders arise from globules by the development of a differentiated pellicle, and themselves give rise to vessels by arrangement in rows and confluence of their cavities. Heusinger generalizes freely and gives little precise information: his style is reminiscent of Naturphilosophie.

Milne Edwards (1823) was strongly influenced by Prochaska, Fontana, and Wenzel. He made a systematic study of the microscopical structure of many organs. He found 'globules' in the connective tissue of man and various animals, in the peritoneum, conjunctiva, the mucous membrane of the intestine, voluntary muscle, tendon, skin, the walls of arteries and veins, and in the white and grey matter of the brain. In nearly every case he notes that the globules are 1/300 mm. in diameter. Unfortunately he gives no figures, and it is impossible to guess exactly what he saw. In some cases he may have been looking at nuclei, in others at lipoidal droplets, in others again he may have seen cells; but if he did, it is difficult to account for their uniformly spherical shape and minute and unvarying size.

The last in the direct succession of the globulists was Dutrochet (1824). Himself mainly a botanist, he relied to a large extent on Milne Edwards for his information about the microscopical structure of animals; but he claims (p. 201) to have verified the latter's observations. For him, 'all the organs of animals are composed of agglomerated globular corpuscles' (p. 13; see also pp. 200–1).

Although there was some truth in the claims of the globulists, and although they did pave the way for a true understanding of the microscopical structure of animals, yet some check to their errors was urgently needed. It was provided by Hodgkin and Lister (1827). Using the improved microscope designed by Lister, they found no globules, but only fibres, in striated muscle and in the muscle of arteries. They looked in vain for globules in nerves. They saw no globules in brain, but only very small particles, which they regarded as resulting from the disintegration of the tissue. They saw no globules in connective tissue (or 'cellular membrane', as they called it). They found human red blood corpuscles to be concave, and the particles of pus to be irregular in shape. They found globules only in milk. They were aware that their results differed from those of Milne Edwards, who was a friend of Hodgkin, and attributed the difference to the imperfection of Edwards's microscope. There can, indeed, be little doubt that many of the globules reported by the early microscopists were images of minute particles, smaller than any ordinary cells, but surrounded by haloes. The fact that the excesses of the globulists were exposed by Lister's microscope seems significant; for the particular advantage of his instrument was that spherical aberration was corrected and the 'ring' appearance round small particles thus reduced. His objectives, though not perfected by this time, must already have been good. The work of Hodgkin and Lister was a healthy and much-needed corrective. They were supported by Grainger (1829), whose own observations agreed with theirs.

The time had now arrived when microscopists were beginning to see actual cells in various animal tissues. Von Baer (1828, pp. 144–5) noticed that the elements of which an embryo consists—fibres, globules, and platelets—become smaller as development proceeds. He uses the words Kügelchen and Körnchen interchangeably. He says of the Körnchen in the developing chick that they 'are so large, in relation to the parts that they compose, that one might say that the embryo at a very early stage resembles a picture made of paving-stones or blocks of granite. On the first day the notochord consists almost entirely of one row of such globules (Kügelchen), which one can count with tolerable accuracy . . . the individual globules can be distinguished in the embryonic area with moderate magnification; the embryo appears to contain several hundreds of them.' There is no reason to doubt that these Körnchen and Kügelchen were cells.

Dutrochet (1837) illustrated a small fragment of the brain of the frog as seen under the microscope. The figure (Fig. 3 on Plate 30) shows a large number of cells and a small vessel running among them. They were probably nerve-cells, though the figure does not permit this to be concluded with certainty. It must be allowed that Dutrochet was not an exponent of animal histology. He remarks (p. 470) that 'when observing with the microscope the tissue of the brain, the liver, the kidneys, the spleen, etc., in a frog, for instance, one really notices no difference'. Purkinje (1838) found that the most diverse organs consist of Körner, Körnchen, or Cylinderchen, often associated with fibres. He found this to be true of the glands of the mammalian stomach, the liver, the salivary glands, the pancreas, unspecified mucous glands, the ear-wax glands, kidneys, testes, epididymides, epidermis, mucous epithelia, ciliated membranes of the respiratory tract and of the female genital system, spleen, thymus, thyroid, and lymph-glands. His paper has a more modern aspect than those of the globulists, and there is reason to believe that he saw the cells of most or all of these organs. The Cylinderchen were the cells of columnar epithelium. By this time, however, the nucleus was beginning to be identified in various tissues. This provided a criterion by which a cell could be recognized as such. The subject now falls within the scope of Proposition II, and will be discussed further under that heading in the second part of this series of papers.

Early Comparisons of Plant and Animal Cells

When considering plant tissues, most of the early observers concentrated their attention upon the cell-walls, which they thought to form a continuous meshwork. Either this meshwork corresponded to the fibres of connective tissue, as Lamarck and the others supposed, or else it seemed to have no counterpart at all in the animal kingdom, in which the tissues consisted largely of 'globules'. Microscopists were slow to realize that the utricles held in the meshes of plant cell-walls might correspond to the globules.

The purpose of this section of the paper is to give some early examples of cases in which actual cells of plants and animals were compared.

Oken's generalizations on this subject have already been quoted (p. 118). It is impossible to be sure that he saw the cells of animals.

Dutrochet (1824, pp. 14–15) said that there is a 'similitude évidente' between the microscopical structure of the brain of gastropods on the one hand and of the pith of *Mimosa pudica* on the other. Raspail (1833, pp. 187, 191) compared the microscopical structure of fat with that of plant tissue. He speaks of 'the analogy of this animal cellular tissue with vegetable cellular tissue'. Valentin (1835, p. 287) described the mesoderm (Gefässblatt) of the chick embryo as composed of large Kugeln, so tightly crowded together 'that they are flattened at many points of contact and often, as [in] the cellular tissue of plants, assume a hexagonal form'. Valentin (pp. 209–10) also refers to a condition resembling the cellular tissue of plants in the ossifying cartilage of the labyrinth of the ear, but it is doubtful whether he is here referring to cells. Müller (1835, p. 25) wrote as follows of the notochord of *Myxine glutinosa*, as seen in transverse section under the microscope: 'The cells are irregular, and unlike one another, but resemble the cells of plants to some extent in that the walls seem to be closed on all sides and mostly touch one another in straight lines, so that irregularly polygonal figures appear in transverse sections.' Müller shows this in Fig. 1 of Plate IX.

Turpin (1837) was led to compare the cells of plants and animals when he undertook a critique of the microscopical studies of a certain Dr. Donné on the liquids secreted and excreted by organic tissues. To check the accuracy of Donné's statements, Turpin repeated most of the observations. Donné had described what were evidently squamous epithelial cells of the human vagina. Turpin says (p. 210): 'After having thoroughly studied the vesicles forming the layer of mucus produced by the vaginal mucous membrane, one cannot avoid seeing in it a cellular tissue that is well-organized and composed, like all vegetable cellular tissues, of an agglomeration by simple contiguity of distinct vesicles living *individually* each on its own account at the expense of the mucous fluid that bathes them on all sides. This animal cellular tissue . . . may be rigorously compared with that of many vegetable cellular tissues.'

Dutrochet reverted to the comparison of plant and animal cells many years after making his first contribution to this subject. He made a direct comparison between the cells of plants and those of the salivary gland of *Helix* (1837, pp. 469–70). 'One sees from that', he remarks, 'that nature possesses a uniform plan for the intimate structure of organised beings, both animal and vegetable.'

As a result of the extensive histological researches that have already been mentioned, Purkinje (1838, p. 175) also drew a comparison of plant and animal cells. 'Consequently,' he wrote, 'the animal organism almost completely reduces itself to three main elementary forms: the fluid, the granular (körnige), and the fibrous. The granular ground-form suggests again an analogy with the plant, which, as is well known, is almost entirely composed of granules or cells (Körnern oder Zellen).'

Comment

The following is the essence of the ideas briefly summarized in the first proposition:

The tissues of most organisms, when examined under the microscope, are seen not to be perfectly continuous, for there is generally a partitioning by cell-membranes, cell-walls, and intercellular matter of various kinds; and this partitioning leaves much of the material of the organism in the form of more or less separate bodies to which the name *cells* is applied. These cells are usually of relatively simple shape in the less differentiated tissues (spheroids, simple polyhedra, &c.).

The facts recorded in this first paper are mainly of historical interest, for the truth of the first proposition is generally admitted and little in the way of critique is possible. It must be remarked, however, that although the knowledge summarized in the proposition was fundamental for the establishment of the cell-theory, yet those who got the knowledge were not at the time in a position to envisage the large superstructure that would eventually be built on their foundations. The first necessary advance was the production of evidence that the various objects that were called cells had in fact important characters in common that made it proper to include them all under a single name. That is the subject of the second proposition which will be considered in the second of this series of papers.

REFERENCES

AWERINZEW, S., 1910. Biol. Centralbl., **30**, 465.

v. BAER, K. E., 1828. *Über Entwickelungsgeschichte der Thiere. Beobachtung und Reflexion.* Erster Theil. Königsberg (Bornträger).

BAITSELL, G. A., 1940. Amer. Nat., **74**, 5.

BICHAT, X., 1812. *Anatomie générale, appliquée à la physiologie et à la médecine.* New edit. Vol. 1. Paris (Brosson, also Gabon).

BOURNE, G. C., 1895. Quart. J. micr. Sci., **38**, 137.

—— 1896*a*. Sci. Prog., **5**, 94.

—— 1896*b*. Ibid., 227.

—— 1896*c*. Ibid., 304.

BURNETT, W. J., 1853. Trans. Amer. Med. Assoc., **6**, 645.

CONKLIN, E. G., 1939. Amer. Nat., **73**, 538.

DOBELL, C. C., 1911. Arch. Protistenk., **23**, 269.

DUTROCHET, M. H., 1824. *Recherches anatomiques et physiologiques sur la structure intime des animaux et des végétaux, et sur leur motilité.* Paris (Baillière).

—— 1837. *Mémoires pour servir à l'histoire anatomique et physiologique des végétaux et des animaux.* 2 vols. and atlas. Paris (Baillière).

E(VELYN), J., 1664. *Sylva, or a Discourse on Forest-trees, and the Propagation of Timber in His Majesties Dominions.* London (Martyn and Allestry).

FONTANA, F., 1781. *Sur les poisons et sur le corps animal.* 2 vols. Paris (Nyon); London (Emsley).

FOSTER, M., 1901. *Lectures on the History of Physiology during the Sixteenth, Seventeenth and Eighteenth Centuries.* Cambridge (University Press).

FRÉDÉRICQ, L., 1884. *Théodore Schwann: sa vie et ses travaux.* Liége (Desoer).

FULINSKI, B., 1916. Zool. Anz., **47**, 381.

GEROULD, J. H., 1922. Sci. Mon., **14**, 268.

GREW, N., 1672. *The Anatomy of Vegetables begun. With a General Account of Vegetation founded thereon.* London (Hickman).

—— 1682. *The Anatomy of Plants. With an Idea of a Philosophical History of Plants. And Several Other Lectures, read before the Royal Society.* Publ. by the author (place not stated).

HALLER, A., 1757. *Elementa physiologiæ corporis humani.* Lausannæ (Bousquet).

—— 1779. *First Lines of Physiology.* Transl. from the correct Latin edition. Edinburgh (Elliot).

HERTWIG, O., 1893. *Die Zelle und die Gewebe. Grundzüge der allgemeinen Anatomie und Physiologie.* Jena (Fischer).

HEUSINGER, C. F., 1822. *System der Histologie.* Eisenach (Bärecke).

HEWSON, W., 1777. *Experimental Enquiries: Part the Third.* Edit. by M. Falconar. London (Longman).

HODGKIN, —, and LISTER, J. J., 1827. Phil. Mag., **2**, 130.

HOME, E., 1818. Phil. Trans., **108**, 172.

—— 1821. Phil. Trans., **111**, 25.

HOOKE, R., 1665. *Micrographia: or Some Physiological Descriptions of Minute Bodies made by Magnifying Glasses. With Observations and Inquiries thereupon.* London (Martyn & Allestry).

—— 1678. *Lectures and Collections made by Robert Hooke, Secretary of the Royal Society.* London (Martyn).

KARLING, J. S., 1939. Amer. Nat., **73**, 517.

LAMARCK, J.-B.-P.-A., 1809. *Philosophie zoologique, ou exposition des considérations relatives à l'histoire naturelle des animaux.* Paris (Dentu, and from the author).

LEEWENHOECK [*sic*], 1674. Phil. Trans., **9**, 23.

LEEUWENHOEK, A., 1686. Phil. Trans., **15**, 883.

—— 1719. *Epistolæ physiologicæ super compluribus naturæ arcanis.* (Vol. 2 of the *Opera omnia.*) Delphis (Beman).

—— 1722. *Opera omnia.* Vol. 1 of Editio novissima. Lugduni Batavorum (Langerak).

MALPIGHI, M., 1675. *Anatome plantarum. Cui subjungitur appendix, iteratas & auctas ejusdem authoris de ovo incubato observationes continens.* London (Royal Society).

—— 1686. *Opera omnia, figuris elegantissimis in æs incisis illustrata.* 2 vols. Londini (Sawbridge).

MARK, E. L., 1879–80. Bull. Mus. Comp. Zoöl. Harvard, **6**, 173.

MECKEL, J. F., 1815. *Handbuch der menschlichen Anatomie.* Erster Band. Halle and Berlin (Buchhandlungen des Hallischen Waisenhauses).

MIALL, L. C., 1911. *History of Biology.* London (Watts).

—— 1912. *The Early Naturalists: Their Lives and Work (1530–1789).* London (Macmillan).

MILNE EDWARDS, H., 1823. *Mémoire sur la structure élémentaire des principaux tissus organiques des animaux.* Paris (Migneret).

M'KENDRICK, J. G., 1888. Proc. Phil. Soc. Glasgow, **19**, 71.

MOLDENHAWER, J. J. P., 1812. *Beyträge zur Anatomie der Pflanzen.* Kiel (Wäser).

MONRO, A., 1726. *The Anatomy of the Humane Bones.* Edinburgh (Ruddiman).

MÜLLER, J., 1834. *Handbuch der Physiologie des Menschen für Vorlesungen.* Erster Band. Coblenz (Hölscher).

—— 1835. *Vergleichende Anatomie der Myxinoiden, der Cyclostomen mit durchbohrtem Gaumen.* Erster Theil. Berlin (Druckerei der Königlichen Akademie der Wissenschaften).

OKEN, (L.), 1805. *Die Zeugung.* Bamberg and Wirzburg (Goebhardt).

—— 1809. *Lehrbuch der Naturphilosophie.* Erster und zweiter Theil. Jena (Frommann).

—— 1843. Ibid. 3rd edit. Zürich (Schulthesz).

POLLENDER, A., 1868. *Wem gebührt die Priorität in der Anatomie der Pflanzen dem Grew oder dem Malpighi?* Bonn (Georgi).

PROCHASKA, G., 1779. *De structura nervorum. Tractatus anatomicus tabulis aeneis illustratus.* Vindobonae (Graeffer).
PURKINJE, —, 1838. Ber. Versamm. deut. Naturf. Aerzte, **15**, 174.
RÁDL, E., 1930. *The History of Biological Theories.* Oxford (University Press).
RASPAIL, F. V., 1833. *Nouveau système de chimie organique, fondé sur des méthodes nouvelles d'observation.* Paris (Baillière).
REMAK, R., 1855. *Untersuchungen über die Entwickelung der Wirbelthiere.* Berlin (Reimer).
RIES, E., 1943. Fort. Zool., **7**, 1.
SACHS, J. v., 1890. *History of Botany* (1530–1860). Transl. by H. E. F. Garnsey. Oxford (Clarendon Press).
SCHLEIDEN, J. M. [*sic*], 1849. *Principles of scientific botany: or, botany as an inductive science.* Transl. by E. Lankester. London (Longman, Brown, Green & Longmans).
—— M. J., 1838. Arch. Anat. Physiol. wiss. Med. (no vol. number), 137.
SCHWANN, T., 1838*a*. Neue Not. Geb. Nat. Heilk. (Froriep), **5**, col. 33.
—— 1838*b*. Ibid. 225.
—— 1839*a*. *Mikroskopische Untersuchungen über die Uebereinstimmung in der Struktur und dem Wachstum der Thiere und Pflanzen.* Berlin (Sander'schen Buchhandlung).
—— 1839*b*. (Contribution to Wagner, 1839.)
—— 1847. *Microscopical Researches into the Accordance in the Structure and Growth of Animals and Plants.* Transl. by H. Smith. London (Sydenham Society).
—— 1884. (Original words of Schwann, not known to be printed elsewhere, given by Frédéricq (1884).)
SEDGWICK, A., 1894. Quart. J. micr. Sci., **37**, 87.
—— 1895. Ibid., **38**, 331.
STIRLING, W., 1902. *Some Apostles of Physiology.* London (Waterlow).
STRASBURGER, E., 1880. *Zellbildung und Zelltheilung.* 3rd edit. Jena (Fischer).
SWAMMERDAM, J., 1737–8. *Biblia naturae; sive historia insectorum, in classes certas redacta.* With a preface by H. Boerhaave. Leydae (Severinum, Vander, Vander).
—— 1758. *The Book of Nature; or, the History of Insects: Reduced to Distinct Classes.* Transl. from the Dutch and Latin original edition by T. Flloyd. London (Seyffert).
TREVIRANUS, G. R. and L. C., 1816. *Vermischte Schriften: anatomischen und physiologischen Inhalts.* 1 Band. Göttingen (Röwer).
TURNER, W., 1890*a*. Nature, **43**, 10.
—— 1890*b*. Ibid., 31.
TURPIN, —, 1837. Ann. sci. nat., **7**, 207.
TYSON, J., 1870. *The Cell Doctrine: Its History and Present State.* Philadelphia (Lindsay & Blakiston).
—— 1878. Ibid. 2nd edit. Philadelphia (Lindsay & Blakiston).
VALENTIN, G., 1835. *Handbuch der Entwickelungsgeschichte des Menschen mit vergleichender Rücksicht der Entwickelung der Säugethiere und Vögel.* Berlin (Rücker).
VIRCHOW, R., 1859. *Die Cellularpathologie in ihrer Begründung auf physiologische und pathologische Gewebelehre.* Berlin (Hirschwald).
WAGNER, R., 1839. *Lehrbuch der Physiologie für akademische Vorlesungen und mit besonderer Rücksicht auf das Bedürfniss der Aerzte.* Leipsig (Voss).
WALDEYER, W., 1888. Arch. mikr. Anat., **32**, 1.
WEISS, P., 1940. Amer. Nat., **74**, 34.
WENZEL, J. and C., 1812. *De penitiori structura cerebri hominis et brutorum.* Tubingae (Cottam).
WHITMAN, C. O., 1893. J. Morph., **8**, 639.
WILSON, J. W., 1944. Isis, **35**, 168.
WOLFF, C. F., 1759. *Theoria generationis.* Halae ad Salam (Hendel).
—— 1774. Ibid. New edit. Halae ad Salam (Hendel).
WOODRUFF, L. L., 1939. Amer. Nat., **73**, 485.

The Cell-Theory: a Restatement, History, and Critique

PART II

BY

JOHN R. BAKER

(From the Department of Zoology and Comparative Anatomy, Oxford)

CONTENTS

PROPOSITION II

Cells have certain definable characters. These characters show that cells (a) *are all of essentially the same nature and* (b) *are* units *of structure.*

THE essence of this proposition can most easily be grasped by considering what would be left of the cell-theory if it were omitted. We should then be in the same position as was Leeuwenhoek (1674), who, having found that a number of tissues consisted of 'globules', was not surprised to find the same structure in milk. This second proposition is concerned with the reasons for supposing that certain objects, called cells, are all to be regarded as strictly comparable with one another and not comparable with globules such as those of milk.

Very gradually, over a period of centuries, it came to be recognized that there is a fundamental living substance, the protoplasm; that this protoplasm commonly occurs in small masses, each provided with a nucleus; and that each of these masses is to some extent separated from its neighbours by a cell-membrane having special characters. Proposition II covers these discoveries and is also concerned with the reasons for supposing that cells are unitary components of organisms and that one cell corresponds with one other cell and not with several. The present paper deals with the discovery of protoplasm and the nucleus. The discussion of Proposition II will be continued in Part III of this series of papers.

The Discovery of Protoplasm

One of the most fundamental facts about cells is that they contain protoplasm as their characteristic constituent, and that, with some partial exceptions that are mentioned on p. 98, this substance never occurs except in

[Quarterly Journal Microscopical Science, Vol. 90, part 1, March 1949]

cells or in objects formed by the transformation of cells. While allowing that the word protoplasm has no absolutely precise meaning, we must acknowledge that there are many substances of which it can be stated with certainty that they are not protoplasm, and that such substances occur commonly between cells, but never constitute cells; while the substance that exists in cells (and in transformed cells) and is called protoplasm has so many positive characters that it is impossible to suppose that we are lumping together under a single name utterly distinct mixtures of organic compounds. This is not the place to give a list of the physical and chemical properties of the substance; we are concerned here only to trace the history of the idea that the cells of plants and animals have a substance called protoplasm as their characteristic component.

The earliest observations and experiments on this substance were not made in connexion with cells. Trembley (1744) made a careful study of the protoplasm of *Hydra* without ever understanding the cellular nature of these animals. He was investigating the microscopical 'grains' (apparently the carotene-granules and nematocysts) that he had discovered. He noticed (p. 56) that when he had teased up a fragment of the body in a drop of water, some of the 'grains' remained bound together by 'une matière glaireuse' (literally, a substance resembling white-of-egg). Trembley stretched a fragment of the body between the points of a quill pen and saw the glairy substance spin out between them (p. 57). He was able to isolate this substance almost completely from the granules. He attributed the cohesion of the granules to the glairy substance. He was also able to stretch a tentacle and to obtain a microscopical view of the part of it lying between two 'grains' (nematocysts): this part consisted wholly of the glairy substance (pp. 63–4). He notes its *transparency* and *tenacity*, the latter being shown by the resistance of the tentacle against breaking when pulled. He attributes to it also the polyp's powers of contraction and expansion.

Duhamel du Monceau (1758, p. 26), in his study of the cellular tissue of plants, mentions the 'substance vésiculaire, ou cellulaire' that fills, as he says, the meshes of the net (i.e. the spaces enclosed by the cell-walls). He remarks that it contracts on desiccation and that it is sometimes coloured.

The discovery of cyclosis by Corti (1774, pp. 127–200) was to play an important part some three-quarters of a century later in leading microscopists to the opinion that the living substance of plants and animals is essentially similar (see p. 95). At the time, the circulation of protoplasm was only an isolated curiosity. Corti uses the name 'Cara' for the various species of freshwater plants on which his observations were made; these included *Chara* and perhaps also *Nitella*. He uses the name *Cara translucens minor, flexilis* for the species in which he first saw the circulation of granules in the long internodal cells.

Treviranus (1811, pp. 78–95) first saw cyclosis in 1803. His observations on this subject were made on *Hydrodictyon utriculatum* and *Nitella flexilis* (which he calls *Chara*), among other freshwater plants. It is clear that he had no knowledge of Corti's discovery. He would appear to have confused cyclosis with the movement of reproductive cells set free by algae, as observed by other authors.

Treviranus calls the protoplasm of freshwater algae 'Gallert oder organische Materie'; he notices the granules in it and mentions that he has seen them also in the cells of the cellular tissue of plants. Indeed, he found each cell of this cellular tissue 'ganz ähnlich' to a segment (Glied) of a conferva (1811, p. 78).

Brisseau-Mirbel (1815, p. 196) uses Grew's word 'cambium' more or less as we might say 'the protoplasm of meristematic tissues'. He describes it as a colourless mucilage that appears wherever new developments are going to occur. He did not understand that it was partitioned into cells, but considered that though a fluid, it contained the 'linéamens' of new structure. In his text-book of histology, Heusinger (1822, p. 41) uses the expression 'Bildungsgewebe (tela formativa)' roughly in the sense of what we should call protoplasm; but his style is reminiscent of Oken's *Naturphilosophie* and he does not give much precise information.

Dujardin (1835) was led to the study of protoplasm by his doubts as to the correctness of Ehrenberg's opinion that the food-vacuoles of ciliates are stomachs joined by an intestine. He was unable to see any tube joining one vacuole to another and his attention was thus directed to the intervening substance. He says that he would perhaps have abandoned these studies, if he had not solved the problem by the discovery of the properties of 'Sarcode'. 'I propose to give this name', he says (p. 367), 'to what other observers have called a living jelly—this glutinous, transparent substance, insoluble in water, contracting into globular masses, attaching itself to dissecting-needles and allowing itself to be drawn out like mucus; lastly, occurring in all the lower animals interposed between the other elements of structure.' It is remarkable that Dujardin at once seized upon most of the important physical characters of the substance he had just named. Indeed, one could hardly improve upon his description in a short statement, except by providing the numerical data that are available to-day. He found (pp. 367–8) that sarcode decomposes gradually in water; unlike albumen, it does not dissolve, but leaves a feeble, irregularly-granular residue. Potash hastens the decomposition; nitric acid and alcohol coagulate the substance and make it white. It spontaneously produces vacuoles within itself. It refracts light much less than oil does.

Dujardin studied sarcode not only in ciliates, but also in *Fasciola* and *Taenia*, in *Nais*, earthworms and other annelids, and in young larvae of insects. He seems generally to have used exudations from rents in tissues for his metazoan material.

Dujardin did not relate his sarcode to cellular structure. Various microscopists, however, began to make short remarks about the substance lying between the nucleus and the boundary of the cell. Valentin (1836), who studied it in nerve-cells, called it the 'Parenchym'. He said that it was 'for the most part a grey-reddish finely granular substance', though transparent and clear as water in fishes (p. 138). He mentions the 'small, dispersed, separate, round particles' in the cytoplasm of various nerve-cells, and figures them (see especially his Figs. 45 and 49 of Tab. VII). These were almost certainly the vacuoles or spheroids that constitute the basis of the so-called Golgi apparatus, and he should presumably be regarded as the discoverer of this cytoplasmic element.

Schleiden (1838, pp. 143–5) seems to have used the word 'Schleim' in more or less the sense of plant protoplasm; but his attention was so much fixed upon the nucleus and cell-wall, and his ideas on the origin of cells so mistaken, that it is impossible to be sure. Certainly he did not believe his 'Schleim' to be an essential part of the cell, except in so far as the nucleus might be formed of it. He says that it occurs in irregular, granular forms without internal structure, and is stained brownish-yellow or brown by tincture of iodine. He seems to have thought that what we should call the cytoplasm of young cells was a watery fluid *containing* granules of Schleim.

Meyen (1839), like Dujardin, was led to the study of protoplasm by investigating the food-vacuoles of ciliates. Like Dujardin, he denied Ehrenberg's opinion that these animals have stomachs joined by an intestine: there are simply watery vacuoles (Höhlen) in a gelatinous substance. 'The true infusoria', he wrote, 'are bladder-like animals, the cavity of which is filled with a slimy, somewhat gelatinous substance' (p. 75). He mentions (p. 79) that similar vacuoles occur in the 'Schleim' of the cells of plants, particularly in the aquatic filamentous forms, but he is so much interested in the vacuoles that he omits to institute a comparison between the gelatinous substance of infusoria and the Schleim of plants.

Schwann (1839 *a*) added little to knowledge of the living substance. He mentions (p. 12) that the cells of the notochord of the frog-larva contain a colourless, homogeneous, transparent substance, which, he says, does not become cloudy at the temperature of boiling water; and he describes the contents of ganglion-cells (p. 182) as being a finely granular, yellowish substance. He gives some account (p. 45) of the 'strukturlose Substanz' of organisms; but this was the supposed Cytoblastem or substance in which cells originate, not the substance of cells themselves. Schwann states specifically (p. 209) that the substance that comes to surround the nucleus in the developing cell is different from the Cytoblastem. He says little about its characters, however, beyond mentioning that it is sometimes homogeneous and sometimes granular.

The first attempt to generalize about the properties of the living substance of plant and animal cells was made by Purkinje on 16 January 1839 at a meeting of the Silesian Society for National Culture. A report on his address was published the following year (Purkinje, 1840*a*). The intrinsic value of his remarks, and the fact that he used the word 'Protoplasma' for the first time in its scientific sense, make it necessary to reproduce a considerable part of what he said. The word Protoplasma had long been used in religious writings in the sense of the 'first created thing'; but it is a surprising—indeed an astonishing—fact that in introducing the term into science, with a very particular and important meaning, he gives no indication that it was not already in current use in this field. He reserves the word 'Zellen' for cells that have distinct cell-walls, using 'Kügelchen' and 'Körnchen' for those that have not. He uses the word 'Cambium' in the same sense as did Brisseau-Mirbel. He wrote as follows:

'In plant-cells the fluid and solid elements have separated completely in space, the former as the inner, enclosed part, the latter as that which encloses it. In the animal development-centre, on the contrary, both are still present in mutual permeation. The correspondence is most clearly marked in the very earliest stages of development—in the plant in the cambium (in the wider sense), in the animal in the Protoplasma of the embryo. The elementary particles are then jelly-like spheres or granules, which present an intermediate condition between fluidity and solidity. With the advance of development the animal and plant structures now diverge from one another; for the former either tarries longer in the embryonic condition or remains stationary in it throughout life, while in the latter on the contrary the hardening process and the separation of the solid and the fluid progress more rapidly, and come to light first in cell-formation and then in the formation of vessels.'

It will be noticed that although he applies the word Protoplasma only to the substance of the embryonic cells of animals, yet he clearly realizes the correspondence of this substance with that of the adult cells of animals and of the meristematic cells of plants. In the case of the adult plant cell, he regards what we call simply the protoplasm as constituting the fluid part of his Protoplasma, the solid part having separated out as the cell-wall. In another paper, published in the same year, Purkinje (1840*b*) follows up these ideas by claiming that there should be a 'Körnchentheorie' as opposed to the cell-theory of Schwann, since plants and animals originate from simpler elementary granules, which in plants become changed into cells, while in animals they either remain as they were or change into various forms of fibres. He does not use the word Protoplasma in this paper, but the idea of a substance common to plant and animal cells is implicit in what he writes.

Jones (1841) denied, like Dujardin, that the 'internal sacculi' of ciliates are connected by an intestine (pp. 56–8). He states that the lowest animals consist of a 'gelatinous parenchyma' (p. 6). He speaks of a 'semifluid albuminous matter' loosely connecting the green granules of *Hydra* (p. 21).

Kützing (1841) helped to direct attention towards the protoplasmic part of plant cells, but unfortunately used a confusing terminology. He claimed that each cell of a conferva consists of three elementary parts: the outer 'Gelinzelle', the 'Amylidzelle', and the 'Gonidien'. The first, from his description, was clearly the cell-wall. The second was what von Mohl was later to call the Primordialschlauch, that is the layer of protoplasm lining the cell-wall on the inside. He describes the Amylidzelle as being coloured brown by iodine; weak acids, alcohol, and drying cause sudden contraction, which cannot be reversed by soaking in water. Kützing made the mistake of supposing that caustic potash converts this layer into starch. The third elementary part was the granular material enclosed by the Amylidzelle (starch-grains, &c.).

A considerable advance was made by Nägeli (1844), who found (pp. 90–1) that there is a slightly granular, colourless 'Schleimschicht' under the whole of the inner surface of the cell-wall of the fully formed cells of green algae and of some fungi. The chloroplasts and starch-grains are attached to it. The whole of the rest of the cells is filled with a water-clear fluid. Nägeli understood that his Schleimschicht corresponded to the Amylidzelle of Kützing, but he objected to the latter's name, firstly because it is not a Zelle (in the sense of 'box'), and secondly on chemical grounds (p. 96). The Schleimschicht, he found, consists of granular slime, which earlier filled the whole cavity of the cell and now lies just within the cell-wall. Its outer surface is smooth, but towards the interior of the cell it forms rather irregular projections. The name 'cell', he insists, is not suitable for such a structure. The Schleimschicht is coagulated by alcohol, weak acids, and water; these are the properties of nitrogenous plant-slime. It is coloured brown by iodine, and it is not changed into starch by potash, as Kützing had said.

These researches of Nägeli to a large extent forestalled the more famous work of von Mohl, who gave the name of Primordialschlauch, or utriculus primordialis, to the protoplasmic layer that lines the inside of the cell-wall of plants (1844, col. 275). This primordial utricle clearly corresponds to the Amylidzelle of Kützing and the Schleimschicht of Nägeli. Von Mohl's term conveys clearly his realization that the cell-wall is not the primary or fundamental part of the cell. He mentions (col. 276) that when a nucleus is present, it lies in the primordial utricle, generally attached to its inner wall; when the nucleus is centrally situated, it is connected to the primordial utricle by slimy threads. The cell-wall stains blue with iodine, while the primordial utricle stains yellowish-brown.

Two years later von Mohl (1846) reintroduced the word 'Protoplasma'. He was quite obviously unaware that Purkinje had already used the word in the same sense. The importance of von Mohl's remarks on this subject justifies rather a long extract. He remarks (col. 73) that if we study a young plant cell, we never find that it contains a watery cell-sap: a viscous, colourless mass, containing fine granules, is dispersed through the cell and is aggregated especially in the vicinity of the nucleus. He thought that this substance was present before the nucleus appeared. 'As has already been remarked,' he writes (col. 75), 'wherever cells are going to be formed this viscous fluid precedes the first solid structures that indicate the future cells. We must further suppose that the development of structure in this substance is the process that initiates the formation of the new cell. For these reasons there may well be justification if, for the designation of this substance, I propose in the word *Protoplasma* a name based on its physiological function.'

(In this translation I have used the expression 'development of structure' to convey the meaning of von Mohl's word 'Organisation'.)

In a footnote von Mohl mentions that Schleiden used the word 'Schleim' in the same sense. Von Mohl objected to this word because it was already used on the one hand loosely for any substance whatever that is of a slimy consistency, and on the other hand in a restricted sense as a synonym for mucus.

He describes (col. 76) how in young cells the nucleus always lies at the centre, surrounded by protoplasm. He proceeds (cols. 77–8) to an account of the origin of the cell-sap. 'Irregularly distributed spaces form in the protoplasm, which fill themselves with watery sap. . . . The older the cell becomes, the more these spaces filled with watery sap increase in size in comparison with the mass of the protoplasm. As a consequence the spaces that have been described flow together into one another.'

It will be allowed that von Mohl had now arrived at a remarkably exact idea of the general plan of a plant cell.

The next necessary step was the discovery that protoplasm is the fundamental constituent of the cells of animals as well as of plants. It might be thought that since the word had first been applied to animal tissues, this step would have been an easy one; but Purkinje's ideas had not received the recognition that was their due, nor had his word 'Protoplasma' been accepted by students of animal cells. The ground gained by Purkinje required to be recaptured.

To Ecker (1848) is due the recovery of the idea that there might be a fundamental substance common to animals of all grades of structure. His object was to discover what there was in lower animals corresponding to the contractile substance of higher animals. He felt that Dujardin's work on sarcode had been disregarded by most histologists. There had been a mistaken tendency to look for parts corresponding to those of the higher animals in the bodies of the lower. 'The body of the Infusoria . . .', he writes (p. 221), 'consists throughout of a simple, homogeneous, half-fluid, jelly-like substance, in which neither cells nor fibres are perceptible—a substance that is sensitive and contractile and in which the essential properties of the animal body are thus not yet confined to particular tissues.'

Ecker concentrated a good deal of his attention on *Hydra*, in which animal he failed to notice the muscular bases of the epithelial cells. He gave the name 'ungeformte contractile Substanz' to the sarcode of Infusoria and the living material of *Hydra*. He found that both were albuminous, soft, either wholly homogeneous and transparent or finely granular; both contained bladder-like spaces or vacuoles; both were in the highest degree elastic and contractile; both insoluble in water, though altered by it; both soluble in potassium hydroxide but hardened and contracted (so he said) by potassium carbonate (pp. 237–8). He claimed to have traced the development of the true striped muscle of the *Chironomus* larva from a completely homogeneous, fibreless, contractile substance.

Ecker's work was important chiefly for its influence on Cohn (1850), who listed the properties of the contractile substance of animals as described by Dujardin and Ecker and then went on to show that this substance was the same as the protoplasm of plant cells. His words (pp. 663–4) are of the utmost importance for the history of the discovery of protoplasm, and must be quoted in full: '*But all these properties are possessed also by protoplasm, that substance of the plant cell which must be regarded as the chief site of almost all vital activity, but especially of all manifestations of movement inside the cell. Not only does the optical, chemical and physical behaviour of this substance correspond with that of sarcode or the contractile substance* (which I had the opportunity to study in the Infusoria, *Hydra*, and Naids)—in particular, both substances are very rich in nitrogen, *are browned by iodine and contracted by stronger reagents—but also the capacity to form vacuoles is inherent in plant protoplasm at all times. . . .*

'*Hence it follows with all the certainty that can generally be attached to an empirical inference in this province, that the protoplasm of the botanists and the contractile substance and sarcode of the zoologists, if not identical, must then indeed be in a high degree similar formations.*

'Accordingly, from the foregoing point of view, the difference between animals and plants must be put in this way, that in the latter the contractile substance, the primordial utricle, is enclosed within a rigid cellulose membrane, which allows it only an internal mobility, normally expressing itself in the phenomena of circulation and rotation—while in the former this is not so.'

Despite this last paragraph, Cohn did not regard the cell-wall as a fundamental part of the plant cell, for he wrote (pp. 655–6): '*In general I comprehend under the expression "primordial cell" that form of the primordial utricle which assumes the aspect of a cell and appears either altogether devoid of a rigid cell-membrane, or independent and isolated from one.*'

It would be difficult to exaggerate the importance of the contribution to our knowledge of protoplasm made by Cohn in the passage just quoted. The contributions made subsequently by Unger, Schultze, Haeckel, and their contemporaries, were amplifications of ideas first formulated by Cohn.

Von Mohl now devoted a book (1851) to the characteristic features of the anatomy and physiology of the plant cell. He remarks (pp. 42, 44) that the protoplasm constitutes a relatively small part of the fully developed plant cell, owing to the large size of the spaces occupied by the cell-sap, which does not mix with the protoplasm. It is difficult to be certain of the exact meaning attached by von Mohl to his word 'Primordialschlauch'. Did he mean the whole of the protoplasmic layer that lies below the wall of the plant cell, externally to the vacuole? or did he mean only the external membrane of this protoplasmic layer? There are remarks on pp. 41–4 which suggest that he was referring to the cell-membrane in the modern sense; but other passages in his writings do not confirm this view, and he does not figure the cell-membrane separately from the protoplasm. He does, however, give a remarkably good figure of typical plant cells, reproduced here as Text-fig. 1.

Remak (1852, p. 53) now adopted von Mohl's botanical word 'Protoplasma' in referring to the substance of the egg-cell and embryonic cells of animals.

The course of progress was now briefly interrupted by an extraordinary episode. T. H. Huxley (1853) made an attempt to discredit not only the view that protoplasm is the fundamental living substance, but also the cell-theory as a whole. He recognized two constituents of tissues: the endoplast (which we should call the protoplasm) and the periplast (intercellular material). His object was to show that life depends primarily upon the intercellular material. 'So far from being the centre of activity of the vital actions', he writes (p. 306), 'it [the endoplast] would appear much rather to be the less important histological element. The periplast, on the other hand, which has hitherto passed under the names of cell-wall, contents, and intercellular substance, is the subject of all the most important metamorphic processes, whether morphological or chemical, in the animal and in the plant.' The endoplast, he says (p. 312), 'has no influence nor importance in histological metamorphosis.' 'We have tried to show', he says (p. 314), 'that they [the cells] are not instruments but indications—that they are no more the producers of the vital phenomena, than the shells scattered in orderly lines along the sea-beach are the instruments by which the gravitative force of the moon acts upon the ocean. Like these, the cells mark only where the vital tides have been, and how they have acted.'

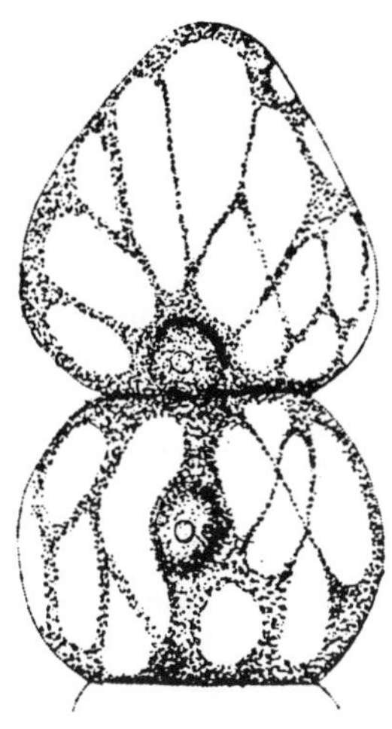

TEXT-FIG. 1. Von Mohl's figure of typical plant cells. The figure represents part of a hair (probably a staminal hair) of *Tradescantia Selowii*. (Von Mohl, 1851, Tab. I, Fig. 7.)

In an important paper Unger (1855) brought strong support to the views that had been formulated five years before by Cohn. After considering the properties of plant protoplasm, and especially its movements, he concludes (p. 282): 'So all this suggests that protoplasm must be regarded not as a fluid, but as a half-fluid contractile substance, which is above all comparable to the sarcode of animals, if indeed it does not coincide in identity with the latter.' Schultze (1858) next described the movement of granules in marine diatoms and compared it with that seen in *Noctiluca* and in the pseudopodia of *Gromia*, Foraminifera, and Radiolaria. In his oft-quoted paper of 1861, which will be considered further in Part III of this series of papers, he mentions that Remak's adoption of von Mohl's word 'Protoplasma' for the substance of animal cells has not been generally copied, and says that he himself will use it henceforth. His example was probably influential.

Schultze's account of the movement of granules in protoplasm was attacked by Reichert (1862*a* and *b*), who claimed that the appearance was illusory. Schultze (1863) had little difficulty in showing that Reichert was mistaken. He proved that the granules of *Gromia* and other rhizopods are real and that they display characteristic movements during life. He also studied the hairs on the stamens of *Tradescantia* and the parts of other plants in which cyclosis is observed. He found a remarkable agreement. 'The movements in the protoplasm of plant cells', he writes (p. 65), 'resemble those in the pseudopodia of the Polythalamia [Foraminifera] so closely, that when the arrangement of the protoplasm is of the kind that occurs, for example, in the cells of the staminal hairs of *Tradescantia*, no difference between the two kinds of movement is to be discovered.' Schultze also showed that chemical and physical influences had similar effects on plant and animal protoplasm.

The pseudopodia of Foraminifera and Radiolaria lent themselves particularly well to studies of protoplasm. Haeckel (1862) made a careful investigation of its properties as revealed in the latter group. He used his powerful influence in support of Cohn and his successors. He wrote (Häckel, 1868, p. 108): 'The *protoplasm* or *sarcode theory*—the doctrine that the albuminous contents of animal and plant cells (or more correctly, their "cell-substance") and the freely motile sarcode of the rhizopods, myxomycetes, protoplasts, etc., are identical and that in both cases this albuminous substance is the originally active substrate of all the phenomena of life—may well be characterized as one of the greatest and most influential achievements of the newer biology.' After paying tribute to the work of Cohn, Unger, and Schultze, he continues (p. 109): 'I have myself striven for a number of years to support and extend this doctrine by numerous observations.'

Meanwhile a strange figure had entered the field. Beale was independent to the point of perversity; he insisted on using a private terminology of his own; and his writings were marred by their polemical character. Had he understood better how to integrate his own discoveries with those of others, he would have made greater contributions to research on protoplasm. Beale first made his ideas public in a series of lectures to the Royal College of Physicians in 1861 (Beale, n.d.). He distinguished between *germinal* and *formed* matter. The former, to which he ascribed the power of infinite growth, evidently corresponds to protoplasm, while the latter is the intercellular material. He regarded an affinity for carmine as a particularly striking character of the germinal matter. Beale's chief interest was in the synthetic function of the germinal matter, and his important contributions to this subject will be discussed under the heading of Proposition IV. The nucleus was for him the quiescent part of the germinal matter. Eventually he accepted the word protoplasm and wrote a book with that name (Beale, 1892); but, wayward to the end, he remarks in it that 'Nowhere in the world is the essential living element a "cell" ' (p. 203).

Brücke (1862) brought a new insight into protoplasmic studies. He insisted (p. 386) that protoplasm has 'Organisation'. He denied (pp. 401–2) that it is either solid or fluid. He objected to its being called a slimy or jelly-like substance, for he thought that this was like the description of a medusa as a gelatinous mass by someone who was ignorant of its organization. The cell-contents must have a complex structure in order to be able to perform the vital activities. His argument was largely deductive and seems to have had little influence on his contemporaries; but it foreshadowed the outlook of a later generation.

Although the Foraminifera and Radiolaria were well adapted for research on the living substance, yet the wholly protoplasmic nature of the Mycetozoa, combined with their fairly large size and ready availability in inland laboratories, made them of predominant importance. In the face of organisms which, in their active, plasmodial phase, contain no 'periplast' whatever, it was impossible any longer to maintain such views as had been put forward by Huxley in 1853. It was this that gave special importance to de Bary's careful study of the group. His description of the protoplasm of Mycetozoa (1864, p. 41) deserves quotation. 'The ground-substance always presents itself as a colourless, translucent, homogeneous mass, exactly similar to the homogeneous contractile substance that is known in the body of amoebae, rhizopods, etc., and is designated as *sarcode, unformed contractile substance*, and latterly, like the component of plant-cells which is in many respects analogous, as *protoplasm*. . . . As for its chemical character, rose-red colouring with sugar and sulphuric acid and with Millon's reagent, together with yellow colouring by iodine, indicate a rich content of albuminous substances. Alcohol and nitric acid cause coagulation; in acetic acid the substance becomes thin (blass) and transparent; in potassium hydroxide solution, even when dilute, it dissolves; the same occurs in potassium carbonate solution, though often after the first action of this reagent has produced a contraction.' It will be agreed that this is a remarkably exact short description of protoplasm.

Kühne (1864) brought strong evidence from many sources of the close similarity of plant and animal protoplasm. He studied the living substance in *Amoeba, Actinophrys, Didymium* (a mycetozoan), in the cells of the connective tissue and cornea of the frog, and in those of the staminal hairs of *Tradescantia*. He observed protoplasmic movements and noted the effects of reagents, of temperature changes, and of the passage of an electric current. He obtained protoplasm from staminal hairs (pp. 100–1) and was so struck by its resemblance to that of *Amoeba* that he called particles of it 'vegetabilische Amoeben'.

Huxley's rhetoric was now to be used once more on the subject of protoplasm. Without giving any indication that he had reversed his opinions or had made any observations or experiments that could cause him to do so, he plunged into powerful support of the protoplasm theory. The occasion was a Sunday evening address given in the Hopetoun Rooms, Edinburgh. According to the careful report given in the *Scotsman* (Huxley, 1868, p. 7), he described protoplasm as 'the bases [*sic*] of physical life'; the expression 'the physical basis of life' first appeared in print as the title of his article in the *Fortnightly Review* (Huxley, 1869), which followed closely the Edinburgh address. 'Beast and fowl,' he wrote (pp. 134–5), 'reptile and fish, mollusk, worm and polype, are all composed of structural units of the same character, namely, masses of protoplasm with a nucleus. . . . What has been said of the animal world is no less true of plants. . . . Protoplasm, single or nucleated, is the formal basis of all life.' The discovery had been made by others: Huxley's contribution was first opposition and then a phrase.

It is to be noticed that the essential similarity between the living matter of plants and animals was discovered by examination of the ground cytoplasm before the existence of mitochondria in both was recognized.

A relatively small point remains. Russow (1884, pp. 578–9) discovered that in the medullary rays of certain plants there exists intercellular material that colours like protoplasm with iodine and dyes. In *Acer* he found thin threads connecting this intercellular with cellular protoplasm. In the same year Fromman (1884) claimed that protoplasm exists in the intercellular spaces of the hypocotyl of *Ricinus*. Like Russow, he said that it reacted to iodine and dyes in the same way as cellular protoplasm. He stated that it often contains single starch grains and small chloroplasts. Intercellular material had already been studied in the cotyledons of the pea by Tangl (1879), who regarded it, however, as secreted matter. The corresponding intercellular substance in the cotyledons of *Lupinus* was seen and figured by Michniewicz (1903), who later saw bridges connecting it with cellular protoplasm (1904). The intercellular material in the cotyledons of *Lupinus* was studied in considerable detail by Kny (1904*a*). He found that reactions for proteins were positive. He concluded from the bleaching of methylene blue solutions and the blue coloration with guaiacol and hydrogen peroxide that the intercellular material respires. Indeed, he found that it showed the same reactions as cytoplasm in all respects, except that studies with proteolytic enzymes suggested that it contained more protein. His general conclusion was that the substance was in fact intercellular protoplasm. In a second paper (Kny, 1904*b*) he showed that it is connected with the cells of the cotyledons by narrow bridges. Like Russow, he found that intercellular protoplasm may contain small starch grains.

This subject would be of considerable importance from the point of view of the cell-theory if it could be shown that intercellular protoplasm ever exists in the absence of any connexion with nucleated cells; but this would not appear to be so. The tissues of animals do not provide any close counterpart to the intercellular protoplasm that occurs occasionally in plants.

The Discovery of the Nucleus

Some of the older botanical writers (e.g. Balfour, 1854) call the nucellus of the ovule the nucleus. This may give rise to misunderstanding. It was stated by Meyen (1839, p. 250), for instance, that both Grew and Malpighi saw the 'Kern' of the 'Eychen' (ovule) of plants. On the same page he uses the word 'Nucleus' as equivalent to 'Kern'. Reference to the relevant parts of Grew's and Malpighi's works shows that there is no question of the object named being a nucleus in the modern sense (see Grew, 1682, p. 210 and Tab. 82; Malpighi, 1687, p. 71, and Fig. 233 on Tab. xxxvii).

Nuclei were in fact first seen by Leeuwenhoek, whose description of them is contained in a letter sent to the Royal Society in 1700 (see Lewenhoek, 1702, p. 556; Leeuwenhoek, 1719, p. 219). The discovery was made in the red blood corpuscles of the salmon. His description of the figure made by his draughtsman (see Text-fig. 2) is as follows:

'Fig. 2 ABCD represents the oval particles of the Blood of a Salmon that weighed above thirty pound.

'AB represents the particles that appeared flat and broad, but did not face the eye directly.

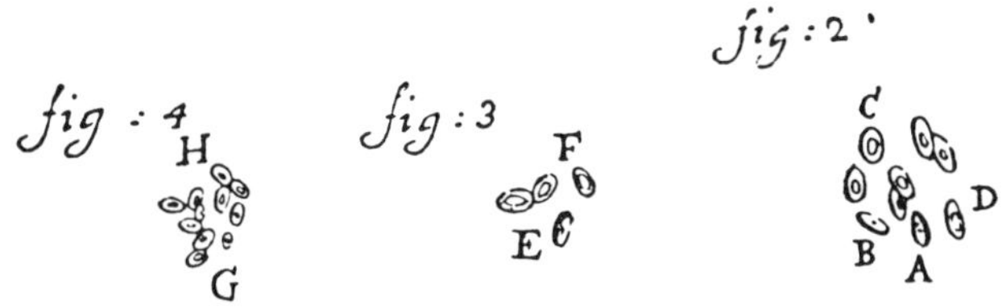

TEXT-FIG. 2. The earliest figure of the nucleus. Red blood corpuscles of the salmon and 'Butt', as represented by Leeuwenhoek's draughtsman. The original numeration of the figures has been retained so that it may correspond with Leeuwenhoek's description as reproduced in the text. (Leeuwenhoek, 1702, Plate opposite p. 220.)

'Those about C came straight upon the eye, and for the most part had a little clear sort of a light in the middle, larger in some than in others, which the Engraver has done his utmost to imitate.'

There can be no doubt that the 'little clear sort of a light' (*lumen* in the Latin version (Leeuwenhoek, 1719)) was the nucleus. Leeuwenhoek also saw nuclei in the blood of a small fish which he calls 'Butt' (1702) or 'Botje'. 'Butt' was a general term in English for flat-fish; in modern Dutch 'bot' is the flounder. Figs. 3 and 4 in Leeuwenhoek's plate (reproduced here as Text-fig. 2) represent the red blood corpuscles of this fish. He refers to the nuclei as 'little shining spots' (1702, p. 557).

Nuclei were seen by some of the early students of Protozoa. Writing to Réaumur in 1744 Trembley gave some sketches of *Stentor* that show the moniliform macronucleus clearly (see Trembley, M., 1943, p. 207). The same observation was made on *Stentor* by Müller (1786, Tab. xxxvi, 8), who calls the macronucleus 'Series punctorum pellucidorum' (p. 262). Müller appears also to have seen the macronucleus in a species of *Colpoda* and a few other ciliates, but there is no indication that he recognized its homology in the different forms. Much later Ehrenberg (1838) saw the macronuclei of many ciliates (*Amphileptus*, *Nassula*, *Chilodonella* ('*Chilodon*'), *Paramecium*, *Spirostomum*, *Stentor*; see his Plates xxiii, xxiv, xxxvi, xxxix), but did not understand their nature. It is convenient to mention Ehrenberg in connexion with this early stage of the history, because his work on the subject was not related to the nuclear research of his own period, for in accordance with his opinion that the ciliates were 'volkommene Organismen', he regarded their macronuclei as male reproductive glands (see, for example, pp. 262, 332, 352).

Hewson (1777) illustrated the nuclei in the red blood corpuscles of many vertebrates—birds, viper, slow-worm, frog, and fishes—and also saw them in the turtle. He made the understandable mistake of suggesting their presence in his figures of the red blood corpuscles of mammals. He was the first to see the nucleus in the blood corpuscle of an invertebrate. 'If one of the legs of a lobster be cut off,' he writes (p. 40), 'and a little of the blood be catched upon a flat glass, and instantly applied to the microscope, it is seen to contain flat vesicles, that are circular like those of the common fish, and have each of them a lesser particle in their centre, as those of other animals.'

Fontana (1781) would appear to have been the first to have seen nuclei in tissue-cells other than those of blood. He made a microscopical study of the slippery substance that coats the skin of eels and described the 'globules' or 'vesicules' contained in it, which were almost certainly epithelial cells derived from the epidermis. 'One saw', he says (p. 276), 'a little body internally, situated in different parts of each globule.' He means that there is not a characteristic position for the little body within the cell. His figures (e.g. Fig. 9 of Plate 1 in Vol. 2) strongly suggest that the 'petit corps' was the nucleus. Indeed, Fig. 10 shows what seems to be a nucleolus within the nucleus. Fontana says, 'The vesicle *a* in Fig. 10 represents one of the vesicles of Fig. 9, in which one observes an oviform body [the nucleus], having a spot (tache) in its interior.' This 'tache' is probably the earliest illustration of a nucleolus. A life-like view was obtained of these cells because they lay in a medium of suitable osmotic pressure. The early observers were accustomed to tease up tissues in water, in preparation for microscopical examination; and it was only when nature provided a suitable medium and thus made the addition of water unnecessary that good views of tissue-cells were obtained.

When Purkinje discovered the germinal vesicle of eggs, he had no means of knowing that there was any correspondence with the 'lumen' seen in the blood corpuscles of fishes or with the 'petit corps' in the epidermal cells of eels. The discovery was not widely known until five years after it had been made. The Faculty of Medicine at Breslau had decided to congratulate Blumenbach in 1825 on the fifteenth anniversary of his taking his doctor's degree, and they wished to send him an original scientific paper to mark the occasion. Purkinje's offer to write a memoir for this purpose was accepted. It was made available to the world at large in 1830 (see Purkinje, 1830 and 1871). The germinal vesicle of the hen's egg was described in this memoir as follows (1830, p. 3):

'Thus the scar [germinal disk] of the ovarian egg contains a special part, peculiar to itself, a vesicle of the shape of a somewhat compressed sphere. This vesicle is limited by a very delicate membrane and filled with a special fluid, perhaps connected with procreation (for which reason I might call it the germinal vesicle); it is sunk into a white breast-shaped projection composed of globules and perforated in the middle.' This remarkable passage contains the earliest mention of the objects later to be named the nuclear membrane and nuclear sap.

Coste (1833, col. 243) showed that the egg of the rabbit contains a vesicle corresponding to that discovered in the hen's egg by Purkinje. He later published a monograph (Coste, n.d.) giving figures of the nucleus of the rabbit's egg. In the description of his Fig. 2 he labels the nucleus 'vesicle analogous to that which Purkinje has demonstrated in birds'. Bernhardt (1834), who knew of Coste's work, found the nucleus, or 'vesicula prolifera' as he called it, in the egg of ruminants and of the rabbit, squirrel, bitch, cat, mole, and bat. He gives figures showing the nucleus in the eggs of several of these, and remarks that with certain precautions even a 'tiro' could see it. He says (p. 27) that it is round or nearly round or oval and has a sharp outline. The contents are fluid.

Meanwhile, nuclei had been discovered in plant cells. Bauer made drawings in 1802 which showed them in the cells of the loose tissue lining the canal of the stigma of the orchid, *Bletia Tankervilliae.* These drawings were unfortunately not published until much later (Bauer, 1830–8; see Tab. VI). Nuclei were probably seen from time to time in plant tissues without anyone guessing that they had any general significance. For instance, Meyen (1830, Plate III) shows what look like nuclei in the pith of the stem of *Ephedra*, though he himself regarded them as consisting of resin-like material. Brown, who knew of Bauer's drawing of *Bletia*, was the first to recognize that the nucleus is of more than sporadic occurrence, and it was he who coined the name by which this part of the cell has been known ever since. His words are as follows (Brown, 1833, pp. 710–11): 'In each cell of the epidermis of a great part of this family [Orchidaceae], especially of those with membranaceous leaves, a single circular areola, generally somewhat more opaque than the membrane of the cell, is observable. . . . This areola, or nucleus of the cell as perhaps it might be termed, is not confined to the epidermis, being also found . . . in many cases in the parenchyma or internal cells of the tissue. . . . I may here remark, that I am aquainted with one case of apparent exception to the nucleus being solitary in each utriculus or cell.' Brown also saw the nucleus in various cells of Liliaceae, Iridaceae, and Commelinaceae, and in a few cases also in the epidermis of dicotyledons.

The nucleolus, which had been recorded by no observer since Fontana, was now discovered by Wagner (1835) in the oocytes of various animals (*Ovis* (Fig. 2 on Tab. VIII), *Salmo*, *Phalangium*, *Anodonta*, *Unio*). He called it the Keimfleck or macula germinativa. The recognition of the nucleolus was important, because it helped in the identification of nuclei.

From 1836 onwards reports came repeatedly of the existence of nuclei in animal cells. Purkinje (1836) announced that the Körnchen (cells) covering the choroid plexus (apparently of man) are each provided with a small

'Körperchen'. Valentin (1836*a*, p. 97) introduced the word nucleus into the literature of animal cytology. Writing of the cells of the epithelium covering the vessels of the choroid plexus of the brain, he says: 'But each of them contains in the middle of its interior a dark, round kernel (Kern), a structure that reminds one of the nucleus that occurs in the plant kingdom in the cells of the epidermis, of the pistil and so forth.' Valentin classifies the epithelia, distinguishing those in which the cells are nucleated from those in which (as he supposes) they are not (p. 96). In a passage of quite extraordinary interest he makes a careful comparison between the egg and the nerve-cell. The latter he calls the formative sphere (Bildungskugel). 'But in what an astonishing way', he exclaims (pp. 196–7), 'does the basic idea of the form of the unfertilized egg correspond with the basic idea of the structure of the formative spheres!' He compares the membrane of these cells with the vitelline membrane, their 'Parenchym' (cytoplasm) with the early yolk, and their nucleus with the germinal vesicle; and he says that a 'Keimfleck' (nucleolus) occurs in both. In a second paper published in the same year (Valentin, 1836*b*), he again uses the word nucleus, stating (p. 143) that every cell without exception in the epithelium of the conjunctiva of man contains one. He mentions also that the nucleus itself here contains 'a perfectly spherical particle'.

The year 1837 saw the publication of a book of the first importance by Henle (1837), who now described nucleated cells in very diverse human tissues, including even the skin of the glans penis. He uses the word 'Cylindri' when referring to cells of columnar epithelium, but elsewhere uses 'Cellulae'. In describing his Fig. 4 he refers to the 'Cellulae nucleatae' of the human conjunctiva. He illustrates the nuclei of the epithelium of the trachea particularly clearly (Fig. 10). He mentions (p. 4) that the nucleus sometimes contains granules. It is not too much to say that this work, with that of Valentin (1836*a*), marks the beginning of an epoch in cytology, the epoch of the *nucleated cell*. Purkinje's name for the nucleus of the egg was, however, not readily relinquished. Siebold (1837) noticed the nuclei in the eggs and blastomeres of nematodes, but called them the 'Keimbläschen' and the 'Purkinjesche Bläschen' in the former case (p. 209) and 'hellen Flecke' in the latter (p. 212). He calls the nucleolus of the egg the 'Keimfleck'. These names, with their counterparts in other languages, persisted long afterwards, even when the homologies of the objects named were well understood.

The year 1838 brings us face to face with Schleiden and Schwann. To assess their significance in the advancement of cytology is a difficult task for the historian, and a task that has often been lightly undertaken. Too much credit has undoubtedly been given to them by some, and a reaction against exaggerated praise has produced a literature of rather superficial belittlement. It is necessary to realize in what field their chief contributions lay. They lay exactly here, in the part of this second proposition that is concerned with the nucleus. If there had been no Schleiden and no Schwann there would have been considerable delay in the general realization by biologists that the possession of a single nucleus is a characteristic feature of most of the cells of animals and plants. Their work, taken together, provided most powerful evidence that there is a correspondence or homology ('Uebereinstimmung', Schwann called it) between the cells of the two kinds of organisms. Their ideas were far from being so original as has often been supposed; for Schleiden had his precursor in Brown, and Schwann in Purkinje, Valentin, and Henle. The three latter had made great advances in animal cytology during a period in which Brown's work on plant cells was scarcely being followed up, and Schleiden's contribution, therefore, represented a more sudden advance than Schwann's; and Schleiden was also ahead of Schwann and communicated his ideas to him in conversation. Indeed, one of Schleiden's most important functions was to act as a stimulus to Schwann: for one can scarcely read the writings of the two men without realizing that Schwann had the greater mind and made much the more massive contribution. These facts stand out even if we deliberately ignore the polemical character of much of Schleiden's writings. It must be allowed that Schleiden had too much influence on Schwann, for the latter took over, without sufficient investigation, his erroneous views as to the origin of cells. That, however, is beside the point for the present: we are here concerned with the great influence of these two men in getting the nucleated cell recognized as the fundamental building-stone of most organisms.

Schwann tells us (1839*a*, p. 8) that during the course of his work on the nerves of the tadpole of the frog he saw the cells and nuclei (Kerne) of the notochord. In his report on the subject (Schwann, 1837) he said nothing about the notochord; but the image of the notochordal cells remained in his mind. 'One day, when I was dining with Mr. Schleiden,' he tells us (Schwann, 1884, p. 25), 'this illustrious botanist pointed out to me the important role that the nucleus plays in the development of plant cells. I at once recalled having seen a similar organ in the cells of the notochord, and in the same instant I grasped the extreme importance that my discovery would have if I succeeded in showing that this nucleus plays the same role in the cells of the notochord as does the nucleus of plants in the development of plant cells.' The two scientists repaired at once to the anatomical institute in Berlin in which Schwann worked. Here they examined together the nuclei of the notochord, and Schleiden recognized the close resemblance to the nuclei of the cells of plants.

Neither had yet published. Schwann was the first to do so, but it will be convenient to begin with Schleiden's contribution, the 'Beiträge zur Phytogenesis' (1838). It is difficult to escape from a sense of disappointment on reading Schleiden's paper. There is nothing about a 'cell-theory' in it; it is solely concerned with plants; and it contains a great deal of error in connexion with the origin of cells, together with much that is of secondary interest. In one respect, however, it was of first-rate importance: the regular occurrence of nuclei in the young cells of phanerogams was here for the first time demonstrated. Schleiden thus focused attention on the nucleus as a characteristic component of the cell. He was also the first to discover the nucleolus of plant cells, without realizing that it corresponded with the 'Keimfleck' already known in both germinal and somatic cells of animals. He calls it a small body (Körper, p. 141), but does not name it. He regards it as 'consistenter' than the rest of the nucleus. He shows it in several figures (see especially Fig. 25 on Tab. III). Beyond all this, Schleiden produced some ideas on cellular individuality, which will be considered under Proposition V.

Schwann's great contribution was his massive array of evidence that there is an 'Uebereinstimmung' between the cells of plants and animals. He himself concentrated upon the latter, relying on the researches of Schleiden for his knowledge of plant cells. We have already seen in the discussion of Proposition I that others had previously suggested such a correspondence between the plant and animal cell; but Schwann was struck with enormous force by the fact that each contains a corresponding object, the nucleus, itself containing a corresponding organelle, the nucleolus, which he called the 'Kernkörperchen'. It seems clear that he reached his conclusions from his own studies of animal cells and from discussion with Schleiden, before he knew of the discovery by Purkinje, Valentin, and Henle that the nucleated cell is a common constituent of animal tissues. He went much farther than these three: he found the nucleated cell to be not simply a common constituent, but the fundamental basis of structure. He founded his 'Zellentheorie' (1839*a*, p. 197) chiefly on his own discoveries. His contributions to the problem of cellular individuality will be mentioned under Proposition V.

In his first cytological paper (1838*a*) Schwann notes the strong resemblance to the cellular tissues of plants shown by notochordal tissue and cartilage, which he had studied in the larvae of the spade-footed 'toad', *Pelobates fuscus*. He mentions the nuclei of both kinds of cells, each containing one or more 'Kernkörperchen'. He regards the embryo as cellular: 'Since therefore the serous and mucous layers of the blastoderm consist of cells and the blood corpuscles are cells, the foundation of all organs that appear later is composed of cells.' He mentions ganglion and pigment-cells and describes the cellular nature of the lens of the eye and of cancerous growths. This paper was published in January 1838. In it Schwann refers to the forthcoming article by Schleiden (the 'Beiträge'), and claims that the latter's statements about the way in which plant cells multiply are applicable also to animals.

Schwann begins his next paper (1838*b*) by stressing the importance of the nucleus in showing the correspondence between animal and plant cells. Most of the observations reported here were made on pig embryos. He records the cellular nature of horny matter, of the lining of the amnion and allantois, the surface of the chorion, the alveoli for the teeth, and the surface of tooth-pulp: all consist of cells with nuclei. He is puzzled, naturally enough, by striated muscle. He finds nuclei in the cells of the kidney, salivary and lachrymal glands, liver, and pancreas. He notes the nucleated cells in connective tissue and considers the white fibres as projections from them. He records the cellular nature of feathers. 'So the whole animal body, like that of plants,' he remarks, 'is thus composed of cells and does not differ fundamentally in its structure from plant tissue.' In his third contribution on this subject, published in April of the same year (1838*c*), he deals with the cellular structure of cartilage, and shows that nail, fat, and unstriated muscle consist of or develop from nucleated cells. The paper ends with an appendix by J. Müller, in which the existence of nucleated cells in pathological growths (osteo-sarcoma, &c.) is recorded. The valid factual part of Schwann's great book, which was published in the following year (1839*a*), consists largely of a reiteration of what he had already made known in these three papers. It contains, however, a considerable amount of interesting theoretical matter, which will be discussed at the appropriate places in future parts of this series of papers.

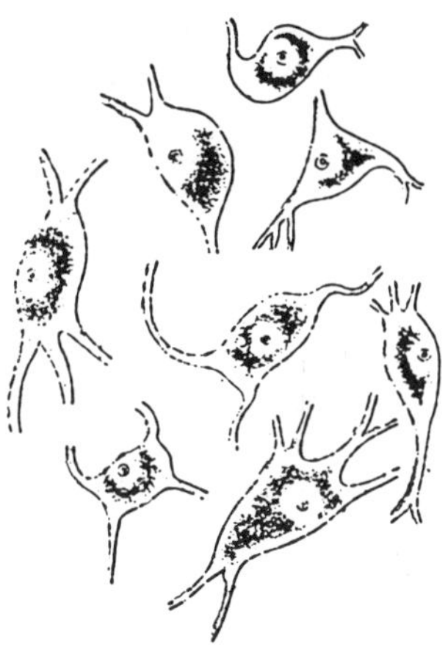

TEXT-FIG. 3. Purkinje's figure of nerve-cells, published in the same year as Schleiden's and Schwann's first publications on cells. The cells are from the black substance of the cerebral peduncle of man. (Purkinje, 1838*b*, Fig. 16 on Plate opposite p. 174.)

Meanwhile, Purkinje, Henle, and Valentin had continued to make discoveries in the same field. Purkinje (1838*a*) mentions the 'Centralkern' in the 'Körnchen' (cells) of the liver. He also gives (1838*b*) an excellent figure of nerve-cells from the black substance of the cerebral peduncle of man. This is reproduced here (Text-fig. 3) so as to give a visual impression of what others were doing in the year in which Schleiden and Schwann made their results known. The nucleus, nucleolus, and pigment are well seen in the figure. Henle followed up his book (1837) with a paper (1838) in which he gave a detailed description of the cellular nature of the epithelia of the human body, including even the lining of the blood-vessels. He is, of course, familiar with the nucleus and the nucleolus; the former he here calls the 'Kern' and the latter, unfortunately, the 'Nucleus'. A useful reminder of the state of knowledge at the time of Schleiden's and Schwann's contributions is provided by the fact that Henle's paper occurs before Schleiden's 'Beiträge' in the same volume of the journal. About the same time Valentin (1838) contributed a curious paper on the differentiation of the cells of the human embryo into their definitive forms. He tried to follow the behaviour of the nuclei (Zellenkerne) during differentiation. What he writes on this subject contains much error, but he was striking out on an important new line and at least he made it clear that the nucleus is far less liable to modification during differentiation than the rest of the cell. Next year Valentin made an unequivocal statement of the correspondence of the nucleated cells of plants and animals (1839*a*, p. 133). He remarks that Schwann has completed the comparison between them.

Valentin now introduced the word 'Nucleolus', in the course of an abstract of Schwann's book (Valentin, 1839*b*, p. 277). He does not mention that he is coining a word, but simply says, 'In mammals the cartilage-corpuscles appear to constitute the whole cell and as such to contain nucleus and nucleolus.' (Turner (1890*a*, p. 11) is wrong in saying that Schwann introduced the word.)

The story of the nucleus may be rounded off by mention of Nägeli's demonstration (1844) that this organelle is a characteristic component of the cells not of phanerogams only, but of all kinds of plants from algae upwards. Thenceforth there was seldom any difficulty in recognizing nuclei; appeal was made especially to the nucleolus as a distinguishing feature in cases of doubt. The use of stains in microtechnique was repeatedly rediscovered during the years 1848–58, as I have told elsewhere (Baker, 1943), and this naturally gave a great impetus to the study of what is usually the most stainable object in cells. Huxley's attempt (1853) to discredit the nucleus is therefore all the more extraordinary. He tells us (p. 297) that Schleiden's belief in the existence of nuclei in all young tissues is 'most certainly incorrect'. The nucleus, he says (p. 298), 'has precisely the same composition as the primordial utricle.' Little attention, however, appears to have been paid to him. From the forties onwards the position of the nucleus in cytology was secure: it was regarded as an essential constituent of the cell. It is to be noted that this conclusion was reached long before there was any general agreement that protoplasm was also a necessary constituent. This may at first seem strange; but it must be recollected that the nucleus is obviously easier to *recognize* than protoplasm, on account of its having morphological as well as physical and chemical characters.

The discussion of Proposition II will be continued in Part III of this series of papers. I thank Prof. A. C. Hardy for his valuable criticism of the typescript of this paper, and Miss O. Wilkinson for conscientious clerical assistance.

REFERENCES

BAKER, J. R., 1943. Journ. Quek. micr. Club, **1**, 256.

BALFOUR, J. H., 1854. *Class Book of Botany: being an Introduction to the Study of the Vegetable Kingdom.* Edinburgh (Black).

BARY, A. DE, 1864. *Die Mycetozoen (Schleimpilze). Ein Beitrag zur Kenntniss der niedersten Organismen.* 2nd edit. Leipzig (Engelmann).

BAUER, F., 1830–8. *Illustrations of Orchidaceous Plants.* With notes and prefatory remarks by J. Lindley. London (Ridgway).

BEALE, L. S., (n.d., *c.* 1861). *On the structure of the simple tissues of the human body.* London (Churchill).

—— 1892. *Protoplasm: Physical Life and Law; or Nature as viewed from without.* 4th edit. London (Harrison).

BERNHARDT, A., 1834. *Symbolae ad ovi mammalium historiam ante praegnationem.* Vratislaviae (Friedlænderi).

BRISSEAU-MIRBEL, C. F., 1815. *Élemens de physiologie végétale et de botanique.* 2 vols. and plates. Paris (Magimel).
BROWN, R., 1833. Trans. Linn. Soc., **16**, 685.
BRÜCKE, E., 1862. Sitzungsber. kais. Akad. Wiss. (Wien), **44**, 2, 381.
BURDACH, K. F. (ed. by), 1837. *Die Physiologie als Erfahrungswissenschaft.* Zweiter Band. 2nd ed. Leipzig (Voss).
COHN, F., 1850. Nov. Act. Acad. Caes. Leop.-Carol., **50**, 605.
CORTI, B., 1774. *Osservazioni microscopiche sulla tremella e sulla circolazione del fluido in una pianta acquajuola.* Lucca (Rocchi).
COSTE, —, 1833. Notiz. Geb. Nat. Heilk. (Froriep), **38**, col. 241.
DUHAMEL DU MONCEAU, —, 1758. *La physique des arbres; où il est traité de l'anatomie des plantes et de l'économie végétale.* 2 vols. Paris (Guerin and Delatour).
DUJARDIN, F., 1835. Ann. des Sci. nat. zool., **4**, 343.
ECKER, A., 1848. Z. wiss. Zool., **1**, 218.
EHRENBERG, D. C. G., 1838. *Die Infusionsthierchen als vollkommene Organismen. Ein Blick in das tiefere organische Leben der Natur.* With Atlas. Leipzig (Voss).
FONTANA, F., 1781. *Sur les poisons et sur le corps animal.* 2 vols. Paris (Nyon); London (Emsley).
FRÉDÉRICQ, L., 1884. *Théodore Schwann: sa vie et ses travaux.* Liége (Desoer).
FROMMAN, C., 1884. Jenaische Zeit., **17**, 951.
GREW, N., 1682. *The Anatomy of Plants. With an idea of a philosophical history of plants. And several other lectures read before the Royal Society.* Place not stated (Rawlins).
HÄCKEL, E., 1868. Jenaische Zeit., **4**, 64.
HAECKEL, E., 1862. *Die Radiolarien. (Rhizopoda radiaria.) Eine Monographie.* Berlin (Reimer).
HENLE, J., 1837. *Symbolae ad anatomiam villorum intestinalium, imprimis eorum epithelii et vasorum lacteorum.* Berlin (Hirschwald).
HENLE, —, 1838. Arch. Anat. Physiol. wiss. Med. (no vol. number), 103.
HEUSINGFR, C. F., 1822. *System der Histologie.* Eisenach (Bärecke).
HEWSON, W., 1777. *Experimental Inquiries: part the third.* Ed. by M. Falconar. London (Longman).
HUXLEY, T. H., 1853. Brit. and For. Med.-Chir. Rev., **12**, 285.
—— 1868. Scotsman, 7886, 7 (9 Nov.).
—— 1869. Fortnightly Review, **5**, 129.
JONES, T. R., 1841. *A General Outline of the Animal Kingdom.* London (van Voorst).
KNY, L., 1904*a*. Ber. deut. Bot. Ges., **22**, 29.
—— 1904*b*. Ibid., **22**, 347.
KÜHNE, W., 1864. *Untersuchungen über das Protoplasma und die Contractilität.* Leipzig (Engelmann).
KÜTZING, —, 1841. Linnaea, **15**, 546.
LEEUWENHOEK, A. A., 1719. *Epistolæ physiologicæ super compluribus naturæ arcanis.* Delphis (Beman).
LEEWENHOECK, —, 1674. Phil. Trans., **9**, 23.
LEWENHOEK, —, 1702. Ibid., **22**, 552.
MALPIGHI, M., 1687. *Opera omnia.* Lugduni Batavorum (Vander).
MEYEN, F. J. F., 1830. *Phytotomie.* Berlin (Haude und Spenerschen Buchhandlung).
—— 1839. *Neues System der Pflanzen-physiologie.* Vol. 3. Berlin (Haude und Spenersche Buchhandlung).
MICHNIEWICZ, A. R., 1903. Sitzungsber. Akad. Wiss. (Wien), **112**, 1, 483.
—— 1904. Öst. bot. Zeit., **54**, 165.
MOHL, H., 1844. Bot. Zeit., **2**, col. 273.
MOHL, H. v., 1846. Ibid., **4**, col. 73.
—— 1851. *Grundzüge der Anatomie und Physiologie der vegetabilischen Zelle.* Braunschweig (Vieweg).
MÜLLER, O. F., 1786. *Animalcula infusoria fluviatilia et marina.* Hauniæ (Moller).
NÄGELI, C., 1844. Z. wiss. Bot., **1**, 1, 34.
PURKINJE, J. E., 1830. *Symbolae ad ovi avium historiam ante incubationem.* Lipsiae (Voss).
PURKINJE, —, 1836. Arch. Anat. Physiol. wiss. Med. (no vol. number), 289.
—— 1838*a*. Ber. Versam. deut. Naturf. Aerzte, **15**, 174.
—— 1838*b*. Ibid., **15**, 178.

PURKINJE, J., 1840*a*. Uebers. Arb. Veränd. schles. Ges. vat. Kult., **16**, 81.
—— 1840*b*. Review of Schwann's 'Microskopische Untersuchungen' (1839). Jahrb. wiss. Kritik, **2**, col. 33.
PURKINJE, J. E., 1871. (Obituary notice.) Proc. Roy. Soc., **19**, p. ix.
REICHERT, C. B., 1862*a*. Arch. Anat. Physiol. wiss. Med. (no vol. number), 86.
—— 1862*b*. Ibid. (no vol. number), 638.
REMAK, R., 1852. Ibid. (no vol. number), 47.
RUSSOW, F., 1884. Sitzungsber. Naturf.-Ges. Univ. Dorpat, **6**, 562.
SCHLEIDEN, M. J., 1838. Arch. Anat. Physiol. wiss. Med. (no vol. number), 137.
SCHULTZE, M., 1858. Ibid. (no vol. number), 330.
—— 1863. *Das Protoplasma der Rhizopoden und der Pflanzenzellen: ein Beitrag zur Theorie der Zelle.* Leipzig (Engelmann).
SCHWANN, T., 1837. Medic. Zeit., **6**, 169.
—— 1838*a*. Neue Not. Geb. Nat. Heilk. (Froriep), **5**, col. 33.
—— 1838*b*. Ibid., **5**, col. 225.
—— 1838*c*. Ibid., **6**, col. 21.
—— 1839*a*. *Microscopische Untersuchungen über die Uebereinstimmung in der Struktur und dem Wachstum der Thiere und Pflanzen.* Berlin (Sander'schen Buchhandlung).
—— 1839*b*. (Contribution to Wagner (1839).)
—— 1884. (Original words of Schwann, not known to be printed elsewhere, given by Frédéricq (1884).)
SIEBOLD, K. T. v., 1837. Article 'Zur Entwickelungsgeschichte der Helminthen' in Burdach, 1837, p. 183.
TANGL, E., 1879–81. Jahrb. wiss. Bot. (Pringsheim), **12**, 170.
TREMBLEY, A., 1744. *Mémoires, pour servir à l'histoire d'un genre de polypes d'eau douce, à bras en forme de cornes.* Leide (Verbeek).
TREMBLEY, M., 1943. *Correspondance inédite entre Réaumur et Abraham Trembley comprenant 113 lettres recueillies et annotées.* Genève (Georg).
TREVIRANUS, L. C., 1811. *Beyträge zur Pflanzenphysiologie.* Göttingen (Dieterich).
TURNER, W., 1890*a*. Nature, **43**, 10.
UNGER, F., 1855. *Anatomie und Physiologie der Pflanzen.* Pest, Wien, und Leipzig (Hartleben).
VALENTIN, G. 1836*a*. Nov. Act. phys.-med. Acad. Leop., **18**, 51.
—— 1836*b*. Repert. Anat. Physiol. (Valentin), **1**, 141.
—— 1838. Ibid., p. 309.
—— 1839*a*. Contribution to Wagner (1839).
—— 1839*b*. Repert. Anat. Physiol., **4**, 1.
WAGNER, R., 1835. Arch. Anat. Physiol. wiss. Med. (no vol. number), 373.
—— 1839. *Lehrbuch der Physiologie für akademische Vorlesungen und mit besonderer Rücksicht auf das Bedürfniss der Aerzte.* Leipzig (Voss).

The Cell-Theory: a Restatement, History, and Critique

Addendum to Part II

BY

JOHN R. BAKER

(From the Department of Zoology and Comparative Anatomy, Oxford)

SINCE writing Part II of this series of papers (Baker, 1949), I have had the good fortune to find a figure, published by Roffredi in 1775, which shows nuclei in the eggs and embryos of a nematode. This figure antedates by fifty-five years Purkinje's description of the germinal vesicle (1830) (or by fifty years, if we use the date of the private circulation of Purkinje's memoir).

Roffredi (1775) describes the anatomy of a small nematode obtained by wrapping flour-paste in linen and leaving the parcel in earth. It is evident from the drawing (Fig. 1 on the plate placed before p. 297) that the animal is a female belonging to the genus *Rhabditis*. The anterior and posterior ovaries and the eggs and young embryos are shown.

Roffredi regarded the whole of the female system as the 'ovaire ou matrice', and called the true ovaries (or ovaries and oviducts) its 'extrémités'. Nuclei are clearly represented in the oocytes that lie in a row along each ovary, and also in the eggs and embryos. Mutual pressure in the ovaries gives each of the oocytes a squarish appearance, and he calls the cell and its nucleus 'le quarré et le globule'. He does not give a written description of the nucleus in the mature egg nor in the embryo.

Spots are shown in some of the nuclei, but there can be no certainty that any of them represent nucleoli, and it seems best to continue to credit Fontana (1781) with the discovery of this organelle.

Roffredi unfortunately failed to distinguish the boundaries of the blastomeres. It is tantalizing to consider how easily he could have discovered cleavage if he had followed up his observations.

REFERENCES

BAKER, J. R., 1949. Quart. J. micr. Sci., **90**, 87.

PURKINJE, J. E., 1830. *Symbolae ad ovi avium historiam ante incubationem.* Lipsiae (Voss).

ROFFREDI, M. D. [*sic*; should be M.], 1775. Journal de physique (Obs. et Mém. sur la physique), **5**, 197.

The Cell-theory: A Restatement, History, and Critique

Part III. The Cell as a Morphological Unit

By JOHN R. BAKER

(*From the Department of Zoology and Comparative Anatomy, Oxford*)

SUMMARY

A long time elapsed after the discovery of cells before they came to be generally regarded as morphological units. As a first step it was necessary to show that the cell-walls of plants were double and that cells could therefore be separated. The earliest advances in this direction were made by Treviranus (1805) and Link (1807).

The idea of a cell was very imperfect, however, so long as attention was concentrated on its wall. The first person who stated clearly that the cell-wall is not a necessary constituent was Leydig (1857). Subsequently the cell came to be regarded as a naked mass of protoplasm with a nucleus, and to this unit the name of *protoplast* was given. The true nature of the limiting membrane of the protoplast was discovered by Overton (1895).

The plasmodesmata or connective strands that sometimes connect cells were probably first seen by Hartig, in sieve-plates (1837). They are best regarded from the point of view of their functions in particular cases. They do not provide evidence for the view that the whole of a multicellular organism is basically a protoplasmic unit.

Two or more nuclei in a continuous mass of protoplasm appear to have been seen for the first time in 1802, by Bauer. That an organism may consist wholly of a syncytium was discovered in 1860, in the Mycetozoa. The syncytial nature of the Siphonales was not revealed until 1879. The existence of syncytia constitutes an exception to the cell-theory. No wholly syncytial plant or animal reaches a high degree of organization.

Natural polyploidy was discovered by Boveri (1887), who was also the first to produce it experimentally (1903). Although many organisms contain some polyploid constituents and others are polyploid throughout their somatic tissues, yet diploid and haploid protoplasts (haplocytes and diplocytes) are the primary components of plants and animals and are still retained as such by most organisms. The haplocyte is more evidently unitary than the diplocyte.

Haplocytes and diplocytes are not composed of lesser homologous units, and with the necessary reservations required by the existence of syncytial and polyploid masses of protoplasm, they may therefore be said to be the fundamental morphological units of organisms.

[Quarterly Journal of Microscopical Science, Vol. 93, part 2, pp. 157-90. June 1952.]

Contents

Introduction

In the first of this series of papers (1948) the cell-theory was restated in the form of seven propositions. The second of these was as follows:

Cells have certain definable characters. These characters show that cells (a) are all of essentially the same nature and (b) are units *of structure.*

Part II of the series (1949) was devoted to the statement labelled (*a*) in this proposition; that is to say, to the fundamental similarity of all cells as revealed by the discovery of protoplasm and the nucleus. We are now concerned with the part of the proposition labelled (*b*); that is, with the idea of the cell as a morphological unit. The idea of the cell as a functional unit cannot escape mention here, but this aspect of our problem will be more fully considered under the heading of Proposition V.

Early Researches Bearing on the Morphological Separateness of Cells

The cellular nature of plants is much more obvious than that of animals, and the earliest cytological observations were naturally made mostly on plants. Since it was the cell-wall and not the cell itself that called attention to the existence of cellular structure, attention was concentrated on the wall as a matter of course. The wall that separates two cells appears single on superficial examination, and the early observers were not inclined to regard cell-walls as separate boxes enclosing material within. The idea of a cell as a morphological unit only originated when it was found that the wall separating two cells was in some cases demonstrably double. The history of this advance will now be related.

It has been mentioned in Part I that Grew described the parenchyma of plants as 'nothing else but a Mass of Bubbles' (1672, p. 79). In his later work he formed a wrong impression and made his well-known comparison with lace. This false simile had a profound influence that remains with us today in such erroneous names as *tissue* (and its counterparts in other languages) and *histology*. We are all accustomed today to speaking of the 'tissues' of the body and may tend to forget that until relatively recent times this word meant nothing else than a textile fabric woven of threads. Until late in the seventeenth century no one could have conceived how such a name could be applied to a part of the body of an organism, and indeed the word was not used in this sense in literary English until about 120 years ago. It is strange to reflect that the modern English usage derives indirectly from a fallacy about the microscopical structure of plants published by Grew in 1682.

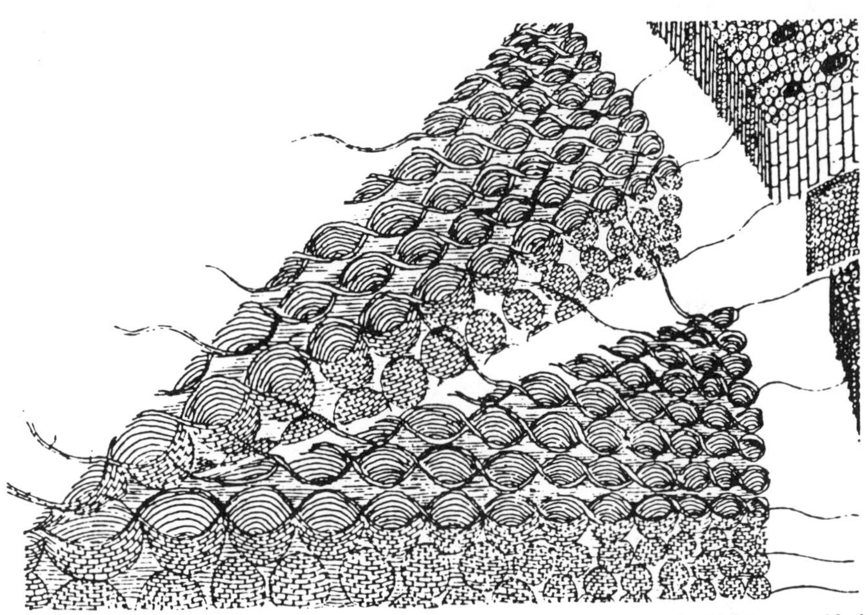

Fig. 1. Part of Grew's drawing of a piece of a branch of the sumach, highly magnified. The supposed fibres, corresponding to the threads of lace, are well shown. (Enlarged from Grew, 1682, plate 40.)

Grew considered that all the parenchymatous parts of a plant, including its fruit, consisted of 'Threds or Fibres', variously woven together. In a well-known passage (1682, pp. 121–2) he compares these fibres to the threads of lace. The comparison is actually to lace while it is being made in a horizontal piece upon a cushion. The pins, inserted vertically into the latter, are imagined to be hollow, and thus represent the vessels of the wood. The lace is supposed to be made in many thousands of layers, one on top of another. The holes between the threads represent the cavities of the cells; and it is understandable that the layers of threads could be added in such a way as to make closed vesicles instead of holes. 'And this', remarks Grew, 'is the true Texture of a *Plant*: and the *general composure*, not only of a *Branch*, but of all other *Parts* from the *Seed* to the *Seed*.'

Of Grew's many figures, the one that illustrates best the ideas just expressed is his representation in perspective of a piece of a branch of the sumach, highly magnified. This figure is here reproduced as fig. 1. It shows clearly his belief in the essentially fibrous nature of plant parenchyma. It follows from what Grew writes and also from this illustration, that if there is a unit in plant tissue, that unit is a fibre and not a cell; for the latter is merely a space left here and there by the intertwining of the fibres.

For well over a century after the time of Grew, attention continued to be focused mainly on the cell-wall rather than on the contents of the cell. The belief that the wall consisted of microscopically-visible fibres did not persist (though the nomenclature based on that mistake was retained); but another kind of error began to be generally accepted. It was thought that the system of cell-walls was a continuous substance throughout the whole of a plant. Spaces (cells and vessels) were supposed to appear in this continuous substance; nourishment was thought to flow through them to the really essential constituent, the continuous membrane or cell-wall. The cells were often regarded as freely open to one another. According to this belief in its extreme form, a plant would consist of two substances, a membranous meshwork and the nourishing fluid filling its meshes; each would be completely continuous. There would indeed be cells, or enlarged meshes; but there would be no unit of structure.

One of the most obstinate adherents to this view was Brisseau-Mirbel, who wrote (1808, pp. 14 and 128): 'The first idea, the fundamental idea is that all vegetable organization is formed by *one and the same membranous tissue*, variously modified. This fact is the base of all the others. The contrary idea is a source of errors. . . . Plants are composed of cells, all the parts of which are continuous among themselves; they present only one and the same membranous tissue.'

In very early times the contrary opinion began to appear, though the development of two opposing schools of thought came slowly, and more than a century was to elapse after the publication of Grew's *Anatomy of Plants* before anyone began to think clearly in terms of the cell as a unit. Grew's contemporary, Malpighi, nevertheless inclined towards what may be called the utricular view: he did not regard the cell as merely a space between interlacing fibres. This appears, for instance, in the account of the microscopical structure of the petals of the tulip and other plants, in his *Anatome plantarum* (1675, pp. 46–47). He remarks that when the surface is torn, the outflowing material contains microscopical objects resembling icicles in shape, which had been entangled loosely together in the intact petal. Each is formed, he tells us, of utricles arranged in a row. One of his illustrations of such a row is reproduced here as fig. 2. There is nothing in such descriptions that would call to mind the lacework of Grew.

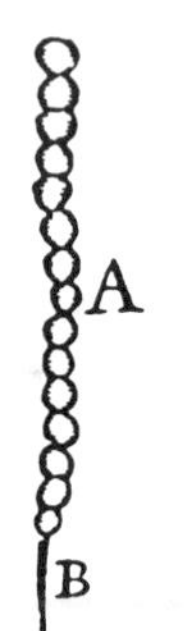

Fig. 2. Malpighi's drawing of a row of cells from the petal of the tulip. (Malpighi, 1675, plate 28, fig. 164.)

The utricular view had its supporters in the succeeding century. Duhamel du Monceau (1758) made some suggestive observations. He separated by maceration small pieces of the 'tissu cellulaire' of the branches of the lime tree. Sometimes he succeeded in detaching little oval bodies of fairly regular shape, which he thought might be the 'vésicules' of Malpighi and Grew; but neither his words nor his illustration (fig. 7 on his plate 2) permit us to feel any more certain of this than he did himself.

Before the cell could be regarded as a unit it was necessary to show that the wall between two contiguous cells was double and that the cell could therefore be isolated as a separate object. This separation can be achieved because of the relative softness or solubility of the pectose or pectic acid of the middle lamella. The first person who clearly demonstrated that plant-cells are separable units was G. R. Treviranus. Referring to the globules mentioned by Wolff (see Part I of this series), Treviranus writes (1805, p. 233), 'I have nowhere seen these little bladders so clearly as in the buds of Ranunculus Ficaria L. A thin section of this, brought under the magnifying-glass in water, allows itself to be divided by the point of a needle into nothing but little bladders (in lauter Bläschen).' He generalizes thus: 'The first beginning of all organization of the living being is an aggregation of little bladders that have no connexion with one another. From these arise all living bodies, just as they are all dissolved into them again.' This statement, which resembles Oken's speculations as expressed in *Die Zeugung* (1805), is of course an induction based on insufficient evidence; but Treviranus's conception of the cell as a unit had actual observation behind it.

Shortly afterwards Link (1807) made a considerable advance towards the understanding of the separateness of cells. He remarks (p. 11) that most authors suppose the existence of an open communication between them, so that the 'Saft' of one can pass into another. Link denies this. When he put cut twigs in coloured fluids, he never observed the passage of the fluid from one cell to another, except when a cell-wall happened to be damaged. He also noticed that certain plant cells have red sap, but are surrounded by others that are uncoloured. To him, cells *and their walls* were separate. His remark on this subject is of particular interest. 'At places where cells adjoin one another', he writes (p. 13), 'one often notices a double line, as if there were a space between the cells.' He illustrates this by a drawing, here reproduced as fig. 3.

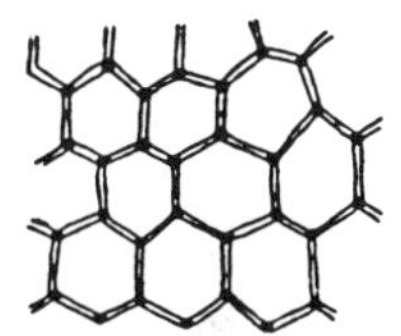

FIG. 3. Link's drawing of a transverse section through the pith of *Datura tatula*, showing the double line where each cell abuts on another. (Link, 1807, plate 1, fig. 2.)

Two years later Link (1809) was quite definite on this subject. 'Cellular tissue', he writes (p. 1), 'consists of little bladders completely separated from one another; but their membranes [cell-walls] usually lie so close to one another that they appear to constitute only a single partition-wall.' He instances the petioles of the large leaves of *Rheum undulatum* and other species as suitable objects for exhibiting the separateness of the cell-walls. The petioles of certain ferns (he mentions *Scolopendrium vulgare* and *Adiantium pedatum*) provide striking examples.

Link continued his investigation (1812) by studying boiled plant tissues, such as kidney-beans and the roots of several garden-plants. He found that not only the cells of the parenchyma but also the elongated bast-cells are separated by this treatment or become separate if gentle pressure is subsequently applied. Link also found completely separate cells, each with its own wall, in many ripe fruits, especially berries. He considered all partitions between cells as originally double; often they remain so and the double wall can be seen, while in other cases it becomes single by subsequent fusion.

Meanwhile L. C. Treviranus (1811) had arrived at similar results. He found that in some cases mere sectioning of plant tissue sufficed to make the cells fall apart, and that this separation can be promoted by pulling gently on the tissue. He reached this conclusion (p. 1): 'It is at once evident to every unprejudiced observer that the cellular tissue of plants is an aggregate of semitransparent bladders, which cohere to a certain extent.' He attacked the contrary opinions of Grew and Brisseau-Mirbel.

Moldenhawer (1812) tried macerating sections of plant tissue in water. He expressed his results very clearly. He remarks on the double nature of the cell-wall.

> But maceration [he writes (pp. 81, 86)], if only employed with due caution, also splits the cellular substance into separate, self-contained utricles . . . [the cellular substance] breaks down sooner or later, according to the firmness of the connexion, into single closed utricles that show no trace whatever of injury, which they necessarily would reveal in the form of irregularly-broken, jagged walls, if there were violent rupture of one and the same continuous tissue. . . . Such an aggregate of single cells has nothing in common with a tissue (Gewebe), and the name *cell-tissue* (Zellgewebe) therefore appears to be less appropriate than that of *cellular substance*, that is, substance consisting of cell-shaped utricles.

Like Link, Moldenhawer mentions cells with differently coloured sap lying close to one another. He explains Grew's 'Threds or Fibres' as mere wrinkles in cell-walls.

Dutrochet (1837) introduced the use of concentrated nitric acid, in a tube immersed in boiling water, as a macerating agent. He showed by this means that the cell-wall is double, and that the substance of plants can be separated into its constituent elements or cells.

By this time, however, the older view had lost its grip: even Brisseau-Mirbel admitted his error. In his work on the liver-wort *Marchantia* he writes of 'the utricular composition of the tissue, which I formerly denied, and of which today I confess the reality' (1835, p. 352). He thus allowed that the cavities of the utricles were not in free communication with one another; but he still adhered to the singleness and continuity of cell-walls. He communicated his paper on the subject to the Académie des Sciences in 1831, but it was not published till four years later. In the meantime his conversion had been complete, and he announced it very frankly in a note appended to the paper.

> The cellular tissue of *Marchantia polymorpha* [he writes (p. 363)] did not offer me spaces between cells. These canals, which are nothing else than the spaces the utricles leave between them, and which for this reason M. Tréviranus calls intercellular, exist in many plants and are absent in others. Thus one can say that the utricles composing cellular tissue are welded together either completely or incompletely. . . . today, when I have obtained the most direct proof of the utricular composition of the tissue, I understand and I see the spaces, which I neither understood nor saw before, and I retract my objections to the fine discovery of M. Tréviranus.

These magnanimous words may be said to mark the end of the controversy about the morphological separateness of ordinary plant-cells.

The Discovery of the Cell-membrane

A clear picture of the nature of the cell could not be obtained so long as the wall was regarded as an essential part. It was necessary to realize that the wall was sometimes present and sometimes absent, while the cell itself was always bounded by a special *membrane*, not mechanically separable from the ground-cytoplasm within. This advance could not be made in one step. It was necessary first to discard the cell-wall as unessential to the idea of a cell, which was then looked upon as consisting of 'naked' protoplasm. The discovery of the cell-membrane came much later.

It must be remarked at the outset that the early workers generally called the cell-wall the *membrane* or *Membran*. This rather confusing usage will appear in various passages quoted in the present paper.

Most cells of animals are obviously devoid of a covering corresponding to the cellulose cell-wall of plants, but one can understand why the cytologists of Schwann's time did not recognize this. It was partly because they were swept away by the new idea of the 'Uebereinstimmung' of all cells, whether plant or animal, and this correspondence would be greatly weakened if one of the most characteristic features of plant cells were found to be absent from those of animals. They therefore looked with confidence for a cell-wall in the animal cell, and in many cases found what they were seeking. The free border of intestinal epithelial cells, the vitelline membrane of eggs, and the cortical layer of ciliates are examples of real structures that seemed to them to represent the cell-wall. Such walls, however, were also described where in fact there are none. This was probably due to the appearance of double lines at the edges of cells, caused by the low numerical aperture of the microscopical objectives available at the time.

The study of animal embryos might have caused a change of opinion, because the blastomeres, which clearly represent the whole organism in its early stages, are devoid of anything resembling the cell-wall of plants; but if blastomeres were to be regarded as cells, as was being suggested, then a cell-wall must be shown to exist, according to the opinion of the day. Reichert (1841), who had the help of du Bois-Reymond, investigated this matter in the developing eggs of amphibians. They isolated uninjured blastomeres in late cleavage-stages and placed them in distilled water under the microscope. According to their account, a surface-membrane was pushed off by endosmosis. They homologized it with the familiar cell-wall and regarded it as evidence for the cellular nature of blastomeres.

Ecker reported that he had seen movements in the blastomere of the frog; he thought these were inconsistent with the presence of a cell-wall, and therefore with the cellular nature of blastomeres (see Remak, 1851). It was Ecker's report that brought the great embryologist, Remak, into the controversy, unfortunately on the wrong side. He opposed Ecker's opinion, and claimed to see two firm membranes surrounding each of the upper blastomeres in the eight-cell stage of the frog; he regarded this as evidence that blastomeres were cells (Remak, 1851). Like Reichert, he saw the membrane surrounding the blastomeres swollen by the osmotic absorption of water, and might have realized the existence of a cell-membrane not corresponding to the cell-wall of plants. He was impressed, however, by the detached envelope that he saw round blastomeres that had been treated with various reagents (hydrochloric, sulphuric, and chromic acids, mercuric chloride, and alcohol), and convinced himself that this envelope corresponded to the cell-wall of plant tissues (1855, pp. 135–6, 173–4). He opposed the view that the surface of a blastomere was merely a modified part of the protoplasm. The cell-walls of plants could be exhibited and distinguished from the underlying protoplasm by simple techniques already in common use in his time, and he looked forward eagerly to the discovery of methods that would play the same part in animal cytology, with equal clarity and certainty.

The first person who stated in unequivocal terms that the cell-wall is not a necessary constituent of the cell was Leydig. He wrote (1857, p. 9):

> . . . not all cells are of bladder-like nature; a membrane separable from the contents is not always distinguishable. For the morphological idea of a cell one requires a more or less soft substance, primitively approaching a sphere in shape, and containing a central body called a kernel (*nucleus*). The cell-substance often hardens to a more or less independent boundary-layer or membrane, and the cell then resolves itself, according to the terminology of scholars, into *membrane*, *cell-contents*, and *kernel*.

The idea that a primitive cell is devoid of a wall was recognized by de Bary in the first of his important contributions to the knowledge of Mycetozoa (1860, p. 161). He wrote of the flagellulae that have emerged from spores: 'In the swarmers there is no cell-membrane in the ordinary sense of the term, but there is indeed a nucleus. As has already been shown above, they are to be regarded as skinless or primordial cells. . . .'

Max Schultze (1860), who had studied protoplasm in various Protozoa, especially the Foraminifera, and had confirmed some of de Bary's work on Mycetozoa, made the following generalization (p. 299): 'But the less perfectly the surface of the protoplasm is hardened to a membrane, the nearer to the primitive *membraneless* condition does the cell find itself, a condition *in which it exhibits only a small naked lump of protoplasm with nucleus. . . .*' Later in the same paper (p. 305) he repeats his definition of a cell as 'ein nacktes Protoplasmaklümpchen mit Kern'.

Schultze now turned to the study of striated muscle and produced the paper (1861) that is so commonly quoted in textbooks of the history of biology. The writers of these, however, have overlooked the rather peculiar character of Schultze's communication. He had set himself the problem of discovering the nature of the 'Muskelkörperchen' or small masses of protoplasm containing nuclei that occur among or outside the contractile elements of striated muscle. He reached the remarkable conclusion that each is to be regarded as a cell. It follows that in the fully differentiated muscle-fibre the contractile elements are extracellular. He knew that each little mass of protoplasm has no cell-wall surrounding it, and his particular point is that this absence of a wall does not indicate that the Muskelkörperchen is not a cell. This led him on to consider what are the essential characters of a cell. He was not content with the definition that had prevailed in the past: 'a vesicular structure with membrane, contents, and nucleus.' To find out what was essential he turned to blastomeres. 'From what has gone before', he wrote (p. 11), 'the component parts of the blastomeres are *nucleus* and *protoplasm*, and our definition of what one has to call a cell assumes the following form: *a cell is a little lump of protoplasm, in the interior of which lies a nucleus.*' The actual words are: 'Eine Zelle ist ein Klümpchen Protoplasma, in dessen Innerem ein Kern liegt.' Schultze refers in a footnote to Leydig's definition. In another place (p. 9) he defines cells as 'little sheathless lumps of protoplasm with nucleus'. He says that the protoplasm holds together because it does not mix with water. 'A membrane', he insists, 'is not necessarily connected with the idea of a cell.' He even considers it a sign of degeneration. 'A cell with a membrane differing chemically from protoplasm is like an encysted infusorium—like an imprisoned monster.' Schultze recognizes that his definition cannot be wholly reconciled with the word *cell*, which conveys the idea of something provided with a distinct wall.

Brücke (1862) agreed with Schultze that the wall is not a necessary attribute of animal cells, but Remak (1862) repeated and amplified his old conclusions. He insisted that the animal cell has at its surface a 'Hülle' or 'Membran', chemically distinct from the ground cytoplasm; and he mentioned once more the separation of this envelope from the underlying protoplasm by chemical agents. He considered that in animals, as in plants, cell-division occurs by the ingrowth of solid septa from the envelope into the protoplasm. Schultze, however, had the support of de Bary in the latter's monograph on Mycetozoa (1864). 'The skin of the cell', wrote de Bary (p. 106), 'is therefore no essential attribute of the cell: it may be formed, but need not.'

A notable advance had been made when the cell came to be regarded as a lump of naked protoplasm with a nucleus, even though the existence of the cell-membrane was not yet clearly recognized. The advance demanded a change in nomenclature, for a cell is a box and a lump of protoplasm is not.

To no one did the old name seem so absurd as to Sachs. Hooke had named the cells of plants after the cells of the comb in a beehive; and if it were right to call the protoplasmic unit a cell, then, according to Sachs (1892, p. 60), a bee should be called a cell, and the cell of the comb should be called the capsule of the cell! Hanstein had been thinking along these lines long before. He recognized that such degree of independence and functional individuality as the cell possesses reside in its protoplasm. This word, however, conveys no sense of an object, but only of the material of which an object may consist. He therefore coined the name 'Protoplast' for the protoplasmic part of a single cell (1880, p. 169). He applied the name to the vital units of both plants and animals. The unit might secrete a wall, but he recognized that this was far from being necessary; for the protoplasts of many animals had no external covering (p. 217). Thus Hanstein agreed in general with Leydig, Schultze, Brücke, and de Bary, but expressed his ideas more exactly. The word protoplast is useful, and modern cytological writings would benefit from a more frequent use of it.

It gradually became apparent that if the vital unit were indeed naked, yet it might at least have a skin. The early comments on this subject are equivocal: we cannot tell whether the authors were referring to the special outer layer or ectoplasm exhibited by certain protoplasts, or to the *membrane* that is always present. Hanstein himself remarked on the skin-like, firmer, 'äussere Hautschicht' of the protoplasm of the plant-cell (1880, p. 167 and fig. 1); he called the Tonoplast the 'innere Hautschicht'. This was a considerable advance beyond the unqualified idea of naked protoplasm, but Hanstein did not undertake the physiological studies that would have been necessary to disclose the real nature of the membrane. The swelling and shrinkage of cells by osmotic pressure might have led to its recognition had not the early work on this subject been confined to the vacuolated plant cell. Pringsheim (1854, p. 51) noticed that solutions of salts, acids, and sugar caused the 'Zellinhalt' to collapse inwards away from the cell-wall. In his famous study of this subject de Vries (1884) defined plasmolysis as the detachment of the living protoplasm from the cell-wall (Zellhaut) through the action of aqueous solutions. He used various plant cells, but chiefly the violet epidermal cells of the lower surface of the leaf of *Tradescantia discolor*. He realized that the membrane responsible for the osmotic phenomena was at the boundary of the vacuole, and since it was involved in the maintenance of turgor, he named it the Tonoplast (1884; see also 1885, p. 469).

Osmotic studies of animal cells would have given new insight into the nature of the boundary of living protoplasm, because the swelling and shrinkage of cells that lacked a vacuole would have directed attention to the true cell-membrane; but such studies were not undertaken at the time, and the view expressed by Leydig and the others prevailed. For more than thirty years it was generally agreed that in its most primitive form the cell consisted of naked ground-cytoplasm (with a nucleus). It required the genius of Overton to recognize a clear distinction between the ground-cytoplasm and cell-membrane, and to bring forward strong evidence that the latter was always present as a covering. He had spent a number of years in a study of the osmotic properties of the living cells of plants and animals. It was known that an 8 per cent. cane-sugar solution caused slight but definite plasmolysis in *Spirogyra*. Overton (1895) tried a solution of ethyl alcohol of the same osmotic pressure, and found no plasmolysis. He thought it possible that this might be due to easy penetration of the outer part of the protoplasm by alcohol. He extended his observations to many diverse plant-cells, with the same result. He then discovered that a number of other substances (various alcohols, ethers, acetone, aniline, phenol) exerted no plasmolytic effect. He realized that it was not the cellulose cell-wall but the 'Grenzschicht' of the protoplasm that was responsible for osmotic effects. He found that in general animal cells resemble those of plants in admitting certain particular kinds of substances and excluding others. On the basis of these observations he produced the diagram of the plant-cell here shown as fig. 4. The diagram makes a clear distinction between ground-cytoplasm, cell-membrane, and cell-wall. Hanstein's diagram (1880, fig. 1) is comparable, but it was not based on a clear understanding of the remarkable properties of the cell-membrane.

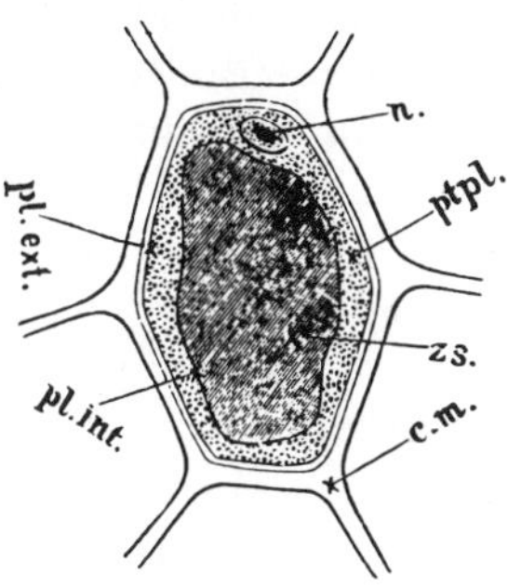

FIG. 4. Overton's diagram of a typical plant-cell, showing the distinction between the cellulose cell-wall (c.m.) and the cell-membrane (pl. ext.). (Overton, 1895, fig. 1.)

Although Overton's diagram marks an important advance, the kind of evidence on which it was based was not quite so satisfactory as an ocular demonstration would have been. A distinction must be drawn between what can be seen and what inferred on indirect evidence. In making his diagram Overton had to decide arbitrarily what thickness he would ascribe to the boundary-layer of the protoplasm.

Overton noticed that the substances that enter cells easily (and hence do not cause plasmolysis) are those that are more soluble in ether, fatty oils, and similar substances than in water, while those that have difficulty in entering cells are those that are readily soluble in water but scarcely or not at all in ether and oil. He was thus driven to the conclusion that the outer layer (Grenzschicht) of the cell must be impregnated with a substance that has dissolving properties similar to those of fatty oils. He rejected the possibility that the substance could be a triglyceride, because he found that filamentous algae could be kept without damage in a solution of sodium bicarbonate that would have saponified a fat. He thought cholesterol or a cholesteryl-ester, with perhaps lecithin and sometimes triglycerides in addition, as the most probable composition of the impregnating substance (Overton, 1899).

So far, Overton had relied on the absence of plasmolysis as evidence that certain substances had entered cells. He now sought more direct indications by following the behaviour of coloured substances (1900). He studied the capacity of various dyes to enter cells, and related it to their solubility in lipoids. He found in general that basic dyes are soluble in melted cholesterol and in solutions of lecithin in organic solvents, and enter living cells easily, while acid dyes are insoluble and do not enter. The exceptions proved the rule, for methylene blue tannate, though basic, is almost insoluble in the solvents mentioned and is not taken up by living cells, while the acid dyes, methyl orange and tropaeolin, are somewhat soluble and are taken up slowly. Overton thus proved the connexion between lipoid-solubility and the capacity of substances to enter living cells, and strengthened the evidence he had previously obtained for the existence of a special lipoid-containing membrane on the surface of cells.

Very various methods have been adopted by later students of the cell-membrane. More refined investigations of permeability have been made; the tension at the surface has been measured; the capacity of various agents to destroy the membrane has been studied; approximations to its thickness have been obtained by indirect methods. The results of these and other experimental studies have been brought together in theoretical diagrams of the molecular and ionic structure of the membrane. The researches on this subject have been admirably reviewed by two active workers in the field (Davson and Danielli, 1943; Danielli, 1951). It suffices for the present purpose to say that although much has been discovered, nothing has occurred to shake the foundations laid by Overton. His work has been of great importance for the cell-theory. Nothing can be a unit that has not a distinguishable boundary. Thanks to Overton we know where that boundary is: we can say with confidence, in many cases, what is part of the protoplast and what is external to it.

Connective Strands Between Cells

There are several theoretical possibilities as to the nature of the connective strands that are sometimes seen to extend from one cell to another. These are illustrated in fig. 5. In A and B the connexion is made solely through the cell-wall. In C the strand consists of the material of the cell-membrane, while in D and E the ground cytoplasm participates, though the membranes, double (D) or fused (E), still to some extent separate the cells. In F there is direct continuity between the ground cytoplasm of the two cells; and though there may be no cyclosis involving the passage of protoplasm from one cell to the other, yet there is obviously an easy route for the diffusion of molecules and ions. It is often impossible to decide which kind of strand is present in particular cases. The subject is difficult from the technical standpoint: fixed preparations are liable to be misleading. As Weiss has pointed out (1940, p. 35), if a preparation shows connexions between cells, we often cannot be sure that they are not formed of coagulated intercellular matter; while if it does not, there may have been connexions in life that were broken by fixation.

Connexions involving the cell-membrane or ground-cytoplasm are often called cell-bridges; but a bridge is something across which there is movement, and it would be begging an important question to give them a general name that implied the transport of material from one cell to another. The non-committal name of Plasmodesmen (singular Plasmodesma), introduced by Strasburger (1901, pp. 503, 607), is preferable. The Greek plural plasmodesmata will be substituted here for the German, as being better suited for international use. It is convenient to employ it in a wide sense, to cover all the possible arrangements shown diagrammatically in fig. 5, C–F. It should be mentioned that the most active study of plasmodesmata took place during

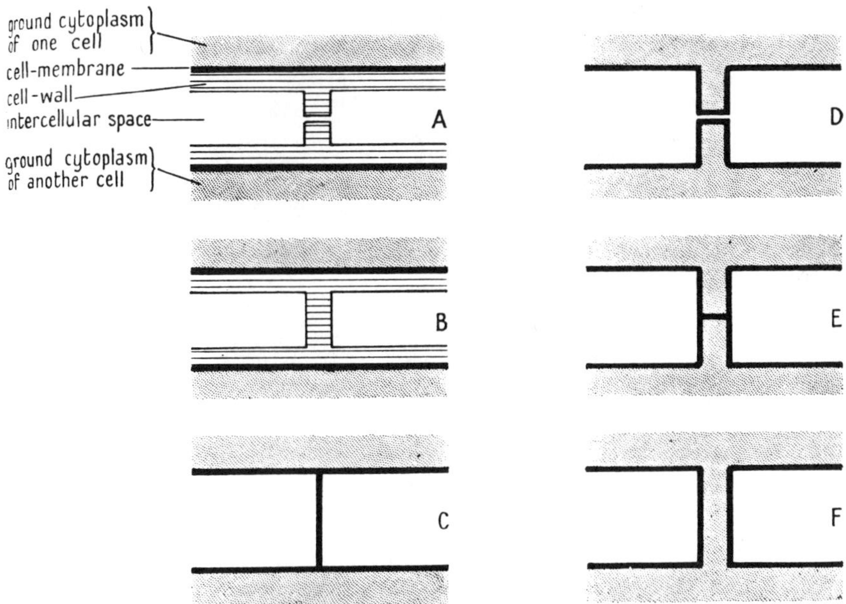

FIG. 5. Theoretical possibilities as to the nature of connective strands between cells.

the long period when cells were commonly regarded as 'naked' lumps of protoplasm, before Overton had proved the existence of the cell-membrane: nice distinctions such as those shown in fig. 5, C–F, would not have had much meaning at that time.

Plasmodesmata must have been seen first in the sieve-plates of plants, which were discovered by Hartig in 1837 (see Esau, 1939) and later re-investigated by him (1854). At first their real nature could not be appreciated because biologists had not yet recognized the nature of protoplasm itself (see Part II of the present series of papers). Sachs, however, who studied them in *Dahlia* (1863), clearly recognized their character, for he mentions that the apertures of the sieve-plates are filled with protoplasm, which he stained with iodine. Modern research shows that two adjacent sieve-tubes are separated by a layer of a substance containing phospholipine, where they abut on one another in the apertures of the sieve-plate; and in some cases there are two layers of the lipoid material, with a very narrow space between (Salmon, 1946, p. 77 and fig. 11). If the lipoid represents a thickened cell-membrane, the plasmodesmata connecting sieve-tubes are of the types represented diagrammatically in fig. 5, D and E.

After the discovery of the plasmodesmata of sieve-tubes, many years elapsed before anything of the kind was shown to exist in other plant-cells. It is stated by Goebel (1926, p. 118) that Hofmeister demonstrated connexions between the cells of the endosperm of *Phytelephas* and *Raphia* in the course of a lecture given during the winter of 1873–4; but nothing was

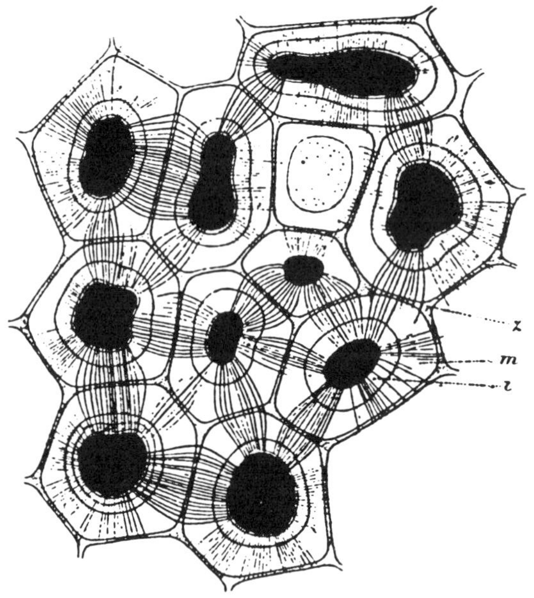

FIG. 6. Tangl's figure showing plasmodesmata in the endosperm of *Strychnos nux-vomica*. The protoplasm (including that of the plasmodesmata) has been stained with iodine-iodide. *i*, inner, and *m*, outer part of the cell-wall; *z*, material lying between adjacent cell-walls. This figure (with those accompanying it in Tangl's paper) was the earliest representation of any plasmodesmata other than the strands connecting sieve-tubes. (Tangl, 1879–81, plate 5, fig. 10.)

published. Fromman (1879), who had come to the conclusion that the protoplasm of the ganglion-cells of the mammalian retina is continuous from cell to cell, turned to plants in search of similar connexions and claimed to find them among the epidermis and parenchyma of the leaves of *Rhododendron* and *Dracaena*. Tangl (1879–81) discovered plasmodesmata between the cells of the endosperm of *Strychnos nux-vomica* and of palms (*Areca oleracea* and *Phoenix dactylifera*). He showed by staining with iodine that the protoplasm was continuous between neighbouring cells through narrow canals traversing the cell-membrane (fig. 6), and considered that the cells entered into a unity of a higher order through these connexions. He thought that there was a striking similarity between the plasmodesmata of endosperm and the (supposed) fibres of the remnant of the mitotic spindle (p. 182). Elsberg (1883) copied the use of silver nitrate and gold chloride from animal histology, and by the use of these reagents claimed to find connexions between various cells of flowering plants. Russow (1884), who had already studied sieve-plates (1881) and was acquainted with Tangl's discovery, announced the existence of intercellular (non-nucleated) protoplasm in the medullary rays of *Acer*, and said that it was connected by strands with cellular protoplasm. He went farther and claimed that 'in every plant throughout its life the whole of the protoplasm stands in continuity' (p. 581). Gardiner (1884) also thought the existence of protoplasmic filaments connecting cells by no means uncommon: the wide perforations in sieve-plates he regarded merely as special cases of a quite usual arrangement. Kienitz-Gerloff (1891*a*) made a general study of the subject, and reported the existence of plasmodesmata in mosses, ferns, horse-tails, and conifers, as well as in mono- and dicotyledons. He came to the conclusion that the whole of the living material of the higher plants is bound together by connexions, though he admitted (p. 23) that in many cases he could not find them. He examined them carefully in the tissues of *Viscum*, where they are particularly evident, and showed that they are not related to the spindle, as this disappears completely before they are formed (1891*b*). Kuhla (1900), like Kienitz-Gerloff, found the mistletoe particularly suitable for investigations of the subject. He considered that all living cells were connected by plasmodesmata. Davis (1905), in a useful general discussion of the whole problem, pointed out that plasmodesmata arise in some cases by incomplete cell-division, in others by outgrowth from previously separate cells. Typical plasmodesmata, in his view, arise in the latter way and are to be regarded as differentiations of the cell-membrane only (pp. 224–6).

Plasmodesmata of particular interest exist in cycads, where they were discovered by Goroschankin (1883), connecting the 'Corpusculum', as he called it, with the surrounding jacket-cells of the endosperm. It is unfortunate that this particular cell appears not to have a generally recognized name; for it cannot properly be called the ovum until its nucleus has divided to form the ovum-nucleus and the ventral-canal-nucleus. It will here be called the developing ovum. Goroschankin found that the cell-wall surrounding it contains a great number of pits, each provided with a sieve-plate through which the protoplasm of the jacket-cells is in open communication with that of the developing ovum. Smith (1904) subsequently discovered the long 'haustoria' connecting the developing ovum of *Zamia*, through pores in its wall, with the jacket-cells. There seems to be no doubt that the haustoria of cycads are nutritive plasmodesmata.

Among the Protista, connective strands between cells are particularly evident in *Volvox*, where they were discovered by Cohn (1875). He described the 'Tüpfelkanäle' or pit-canals in the cell-wall and the delicate thread-like processes that appear to pass through them from one cell to another; he figures these in *V. globator* (in his fig. 1 on plate 2). Strangely enough, Cohn himself thought that there was no actual connexion between the cells, as he supposed that the pit-canals were closed (p. 95); this is curious, because in his figure he shows obvious plasmodesmata connecting the cytoplasm of adjacent cells. Modern studies (Janet, 1912, p. 48) show that *V. globator* provides one of the clearest instances of the kind of connexion that is illustrated diagrammatically in the present paper in fig. 5F. The reality of protoplasmic communication in *Volvox* was recognized by Bütschli (1883). Meyer (1895, 1896) also regarded the strands as 'Plasmaverbindungen'; he pointed out that they are long and thin in *V. aureus*, but shorter and much thicker in *V. globator*. It is particularly suggestive that the growing macrogamete of *Volvox* is abundantly provided with plasmodesmata connecting it with the surrounding cells (Janet, 1912, p. 91); the arrangement is reminiscent of the developing ovum of cycads.

Among the Metazoa, plasmodesmata were first discovered in the skin of mammals. We may overlook Henle's (1841) description of prickles (Stacheln) projecting from the surface of the cells of the chorioid plexus, partly because there is no certainty whether what he saw were cilia or elements of the striated border of the cells, and partly because it is unlikely that these cells are in fact connected by plasmodesmata. The discovery was made by Weber (1858) in suppurating skin and in epithelial cancers. Weber himself, however, regarded the strands as cilia. They were studied by Schultze (1864*a*) in the lower layers of the epithelium of the mammalian tongue and skin. He made the mistake of thinking that each cell had its own prickles (Stacheln), which interdigitated like the bristles of two brushes pressed together. This opinion he retained in a second communication on the same subject (1864*b*). Schrön (1865) considered that the appearances really indicated the existence of canals in a 'Membran' separating skin-cells. Bizzozero (1876), who appears to have waited six years for the publication of his paper, was the first to describe the plasmodesmata of skin correctly as direct connexions between one cell and another. No one has ever produced evidence that they represent connexions of the type shown in fig. 5F. It seems reasonable to regard them as serving to hold the cells together mechanically, while allowing bending movements.

Similar connexions have been reported from time to time in very various animal cells. Heitzmann (1873) was an enthusiastic student of plasmodesmata. He claimed to display them by the use of silver and gold impregnations in a wide variety of mammalian tissues, including bone-marrow and even cartilage. Kultschitzky (1888) claimed to find protoplasmic connexions between the smooth muscle cells of the intestine, and looked forward to the discovery of a universal system of such connexions between neighbouring cells in both plants and animals. Connexions between notochordal cells, resembling the prickles of skin-cells, were reported by von Ebner (1896).

Reports of direct connexions between nerve-cells were much more plausible because of the undoubted transmission of the nervous impulse from one cell to another. Different opinions were held on this subject from early times, but no strong evidence was forthcoming until the eighties. Using his potassium dichromate and silver nitrate method, Golgi (1883, p. 289) never found a single anastomosis among the ramifications of the main processes of nerve-cells; but he considered that other processes existed, which subdivided in a complicated manner and anastomosed, so that the nerve-cells lost their individuality and took part in a nervous reticulum. His (1886) was the first to base the contrary view on sound foundations and state it clearly. As a result of his study of human embryos he came to regard the axons of nerve-fibres as outgrowths from separate nerve-cells that push their way between other tissue-constituents; he denied that these outgrowths ever form actual anastomoses (pp. 509, 513). He remarks: 'I present as an established principle, the proposition *that every nerve-fibre arises as an offshoot from one single cell. This cell is its embryonic (genetisches), nutritive, and functional centre, and other connexions of the fibre are either only indirect, or have originated secondarily*' (p. 513). Shortly afterwards Forel arrived at the same conclusion from his work on the cavy, but expressed it more tentatively: 'I might presume that all fibre-systems and so-called fibre-nets of the nervous system are nothing else than nerve-processes, [each] always [arising] from a particular ganglion-cell' (1887, p. 166). His (1889) went on to describe the outgrowth of the axon from the neuroblast in various vertebrates. The idea of the separateness of nerve-cells has been followed up by many workers, notably Ramón (1934, &c.), in modern times. There are still distinguished neurologists who oppose the neurone-theory and view the nervous system as a reticulum, but some of their arguments are not strong (as, e.g., when Boeke (1940, p. 144) attacks the cell-theory, and the neurone-theory as part of it, on the ground that it 'belonged to the mechanistic and analytic mental attitude' of the nineteenth century). The modern literature referring to the chordates has been well summarized by Nonidez (1944), who reaches the conclusion that nerve-cells are not directly continuous with one another at the synapse.

The nervous system of coelenterates has long been supposed to provide strong evidence for the reticular theory. The belief that the nerve-fibres of these animals are continuous from cell to cell originated with Korotneff (1876), who studied the layer of nerve-cells and fibres that lies below the external epithelium of the acrorhagi ('bourses marginales') of *Actinia*. He described the fibres as running without a break from nerve-cell to nerve-cell. He even thought the fibre maintained its individuality within the nerve-cells. Korotneff mentioned only single rows of cells, but subsequent workers believed that the apparent nerve-net of coelenterates was formed by continuous fibres passing uninterruptedly from cell to cell. This opinion was generally accepted for several decades, though Schäfer had stated in the most definite manner (1878) that each fibre of the bipolar cells of the sub-umbrellar surface of *Aurelia* 'is entirely distinct from, and nowhere structurally continuous with, any other fibre'. He knew that the fibres came into close relation with one another, and thought it reasonable to conclude that nervous impulses passed from one to another, but he considered each nerve-cell with its two processes a separate unit. This was confirmed in modern times by Bozler (1927), who treated the fresh tissues of the jellyfish *Rhizostoma* with reduced methylene blue solution and exhibited the nerve-cells with their processes as separate units (some bipolar, others multipolar); the units make contact with one another but do not anastomose. He denied, therefore, that the nervous system of *Rhizostoma* is a genuine nerve-net. The results of the physiological study by Pantin (1935, *a* and *b*) of the nervous system of the sea-anemone *Calliactis* are consistent with the belief that here also the nervous system consists of cellular units.

There are, of course, cases in which nerve-cells make such evident junctions that an actual syncytium is formed. Thus in cephalopods each of the two first-order giant-cells, which lie in the brain close to the statocyst, sends back a giant axon, and the two axons are connected by a wide bridge as they pass through the palliovisceral ganglion (Young, 1939). Again, the giant fibre that runs along the ventral nerve-cord of the polychaet *Myxicola* is part of a syncytium, for at least 1,300 nerve-cells are continuous with it, directly or indirectly (Nicol, 1948). These interesting facts, however, throw no light on the nature of ordinary synapses. In the light of existing knowledge it is best to draw the provisional conclusion that the nervous system does not provide us with convincing examples of plasmodesmata, though actual syncytia are found in particular cases.

It was found by Berthold that the developing eggs of the nematomorph worm, *Gordius*, are fixed together like grapes in bunches (see von Siebold, 1843). Reinvestigating the matter, Meissner (1856) reported that the eggs originate in groups of 8–20 in the ovary by bulging outwards from a mother-cell, and remain for a long time in organic connexion with one another through their stalks. It is necessary to mention this, because it would have been the first example of plasmodesmata discovered in animals, if true. It appears, however, from the careful work of Vejdovsky (1888, p. 204), that in reality the eggs are only held together in groups by the ovarian epithelium, which becomes very thin in late stages and closely pressed to the surfaces of the eggs.

The connexion of animal eggs with external protoplasm was first reported by von Ihering (1877) in the lamellibranch *Scrobicularia*, six years before Goroschankin's discovery in cycads. According to von Ihering, the follicles of the ovary are lined by a syncytium. In this, an egg develops by the accumulation of protoplasm round a nucleus. The egg then projects into the cavity of the follicle and eventually remains in connexion with the syncytium only by a narrow stalk. Yolk is seen in the syncytium, in the stalk (where it is arranged in regular lines), and in the egg. Korschelt (1886) undertook a detailed study of the morphological relation between nurse-cells and eggs in various insects. Gross (1901) described and figured the long, narrow 'Dotterstränge' that lead from the nutritive end-chamber of the ovarian tubule of Hemiptera to the oocytes; the end-chamber itself is a mass of protoplasm in which nuclei are degenerating or have degenerated. Later (1903) he studied the relation between nurse-cells and oocytes in other insects. In the carabid beetle *Harpalus* he illustrated a nurse-cell nucleus in passage along a narrow neck connecting the oocyte with its food-supply (plate 11, fig. 124). In the hemipteran *Triecphora* he described cords like Dotterstränge connecting some of the nurse-cells with the main protoplasmic mass of the end-chamber, which itself supplies the growing oocytes through long Dotterstränge. This arrangement in Hemiptera was confirmed by Mestschershaya (1931), who investigated the permeability of oocytes and end-chamber to various substances in solution, and reached the conclusion that the end-chamber is adapted to take up food-substances and pass them to the oocyte. There seems to be no doubt that the Dotterstränge of certain insects are genuine plasmodesmata of the type shown in fig. 5F (see, e.g., Korschelt, 1936, fig. 38C).

It was reported by Hammar (1896) that in the cleavage of *Echinus* the outermost protoplasmic layer is continuous from one blastomere to the next. Protoplasmic connexions between blastomeres might be thought to indicate a primary condition, characteristic of cells in general, and the subject attracted attention. Flemming (1896) considered that blastomeres are primarily separate but in some cases become secondarily connected by strands. Andrews reported that in the starfish and other echinoderms the blastomeres send out thin protoplasmic processes that join them together (1897), and that some of the cells of the blastulae are connected in this way (1899). Hammar (1897) made a general study of this subject, using as fixative a saturated solution of mercuric chloride in evaporated sea-water, with the deliberate intention of shrinking the cells and thus increasing the intercellular spaces and exhibiting the strands crossing them. He claimed to find protoplasmic connexions between blastomeres in various invertebrates from coelenterates upwards, but it would be unwise to place much reliance on results obtained by a method so particularly liable to produce artificial appearances. In *Dendrocoelum*, as Fulińksi (1916) showed, not only are the blastomeres not connected by plasmodesmata: they do not even touch one another, but lie separately in a fluid derived from the yolk-syncytium.

The belief of Sedgwick in the continuity of protoplasm from cell to cell has been so influential that it is desirable to treat separately the controversy he aroused. In studying the development of *Peripatus* he noticed that the endoderm cells, previously separate, put out branches that anastomosed (1885). This was the origin of his doubts about the truth of the cell-theory. He was later impressed by the structure of the mesenchyme of elasmobranch embryos, which he described as 'a reticulum of a pale non-staining substance holding nuclei at its nodes' (1894). He regarded the ectoderm and endoderm as 'simply parts of this reticulum in which the meshes are closer and the nuclei more numerous and arranged in layers'. What were taken by others to be sites of cell-proliferation were described by him as places where nuclei multiplied. 'In short, if these facts are generally applicable', he wrote, 'embryonic development can no longer be looked upon as being essentially the formation by fission of a number of units from a single primitive unit, and the coordination and modification of these units into a harmonious whole. But it must rather be regarded as a multiplication of nuclei and a specialization of tracts and vacuoles in a continuous mass of vacuolated protoplasm.' The 'vacuoles' of Sedgwick would be what adherents to the cell-theory would call intercellular spaces not filled by cell-walls or other solid matter. He claimed that nerves were laid down before any trace of nerve-cells could be made out. 'In short, the development of nerves is not an outgrowth of cell-processes from certain central cells, but is a differentiation of a substance which was already in position.'

These arguments were answered by Bourne (1895), who took Sedgwick's objection to the cell-theory to be based essentially on embryological evidence. Bourne himself did not accept the idea that a multicellular organism is actually a 'cell-republic', but he insisted that it is an aggregate of elementary parts, and thought that Sedgwick's views would take us back to the Cytoblastem of Schwann. He considered that in the case of spiral cleavage at any rate the blastomeres are not connected by protoplasmic strands. Sedgwick did not leave Bourne's rather mild criticism unanswered. His study of *Peripatus* had led him to the view that 'the differentiation of the Metazoa had been effected in a continuous multinucleated plasmatic mass, and that the cellular structure had arisen by the special arrangement of the nuclei in reference to the structural changes' (1895). He could not accept the idea of the zygote *dividing* into blastomeres, and he insisted that both ovum and spermatozoon were individuals, simplified so as to make their fusion possible. Animals for him were generally 'tetramorphic': that is, they exhibited four kinds of individual: male, female, spermatozoon, and egg. He thus allowed individuality to the gametes, while denying it to other cells.

Sedgwick's ideas were not novel, for Heitzmann had expressed them long before. In the paper already quoted he wrote (1873, p. 155): '*The animal body as a whole is one lump of protoplasm, in which are embedded to a smaller extent isolated protoplasmic bodies* (wandering bodies, colourless and red blood-corpuscles) *and various other substances that are not alive* (gelatinous and mucous substances in the widest sense, together with fat, pigment granules, &c.).' He compared the whole of a higher animal to an *Amoeba*, and denied that there is any such thing as intercellular substance, even in blood; for him, there was only 'Grundsubstanz' and protoplasm (1873, p. 155). Ten years later, about the time when Sedgwick was first turning his attention to the subject, Heitzmann wrote: 'What have previously been considered as cells prove, in our conception, to be nodal points of a network that traverses the tissues' (1883, p. 57). He applied this generalization to both plants and animals. It was Sedgwick's eminence as a zoologist, however, that gave currency to the new ideas. He spread them not only by his writings but through personal contacts. Dobell, one of the strongest opponents of the cell-theory, was trained in his school.

There is no reason to suppose that protoplasm flows freely through every connexion we may find between cells. Schultze made a comment on this subject long ago that is memorable for its good sense and moderation. 'But

I dispute', he wrote (1861, p. 26), 'that the *individuality* of cell-life is encroached upon by the anastomoses, and I dispute that in normal circumstances, with full integrity of the individual cells, the situation can be even quite roughly interpreted as a protoplasmic vessel-system.' In much the same sense Strasburger (1901, p. 595) drew a distinction between the existence of plasmodesmata on the one hand, and the loss of cellular individuality on the other.

Plasmodesmata should be considered from the functional point of view. Where a necessity has arisen for bulky materials to stream into a particular cell, protoplasmic connexions have evolved for their passage. We have seen examples in the developing ovum of cycads, in the macrogamete of *Volvox*, and in the oocytes of various insects. Where it has been particularly important for cells to combine the property of holding together firmly with that of allowing changes of relative position, strands are seen to pass between neighbouring cells. Flemming (1896) remarked on the physiological necessity for junctions in certain epithelia. Indeed, the mammalian skin provides one of the most familiar examples of plasmodesmata. Another good example is the mesenchyme of the embryos of very diverse animals. In this embryonic connective tissue there are no extracellular fibres, and the function of holding together is served by direct connexions between cells. In some cases we cannot assign a function to the plasmodesmata; but in general we find them serving some particular purpose, and not existing as biological necessities indicative of an essential protoplasmic unity of the whole organism. We have no reason to suppose that cells usually possess them.

Syncytia

The nomenclature of multinucleate masses of protoplasm is very confused. The word *coenocyte* is generally employed in botany to mean a whole plant containing several or many nuclei not marked off from one another by cell-boundaries, but it is also sometimes used for *parts* of plants that contain more than one nucleus in a continuous mass of cytoplasm. The word *syncytium* suffers from having been used in different senses by its originator. Haeckel defined a cell (Cellula) as a lump of protoplasm with a nucleus, and called the expression 'mehrkernige Zelle' a *Contradictio in adjecto* (1866, vol. 1, pp. 275, 296). In coining the word *Syncytium* he restricted its meaning specifically to a complex formed by the fusion of previously separate cells (1872, vol. 1, p. 161); he applied it to the dermal epithelium of calcareous sponges, which he believed to be of this nature. Cienkowsky, however, had invented the word *Plasmodium* for a continuous mass of protoplasm formed by the fusion of previously separate cells (1863*a*, p. 326); he applied the word to a stage in the life-history of Mycetozoa. He knew that this stage was reached by the fusion of nucleate cells, but considered the plasmodium itself to be devoid of nuclei (1863*b*, pp. 435–6). Haeckel accepted this, and thus drew a distinction between a syncytium and a plasmodium (1872, vol. 1, p. 161). Later he supposed that Mycetozoa were at first nucleate and later non-nucleate; he expressed this by saying that they were syncytia that became plasmodia (1878, p. 51). Later again he evidently revised his ideas on the nature of a *Contradictio in adjecto*, for he referred to the Siphonales and other multinucleate organisms that lacked cell-boundaries as polykaryote cells (1894, p. 70). Finally he stated distinctly that a *Bryopsis* or *Caulerpa* consists of a single cell with many nuclei, and gave a new definition of his word syncytium: 'The whole body consists of a single colossal cell, which includes many nuclei in its voluminous body' (1898, vol. 2, p. 421), He applied the name to Siphonales, Mycetozoa, *Actinosphaerium*, and certain Foraminifera, but not to any constituent part of any organism; and he equated syncytia with plasmodia.

Gegenbaur applied the term syncytium to striated muscle (1874, p. 26); Huxley introduced it into English, with Haeckel's original meaning (1877, p. 113). Delage and Hérouard (1896, p. 41) also restricted the sense of the word to cases in which previously separate cells fuse together.

In an attempt to reduce the confusion caused by these various meanings I shall use the word syncytium to mean *any obviously continuous mass of protoplasm that contains more than one nucleus, whether that mass constitutes a whole organism or part of an organism, and whether the bi- or multinucleate condition has been reached by aggregation of previously separate cells, or by nuclear division without cell-division, or by both aggregation and nuclear division.* I shall restrict the word *plasmodium* to *a syncytium formed by aggregation of previously separate cells, whether subsequent cell-division increases the number of nuclei or not.* The word *coenocyte* appears to be superfluous and will not be used. It may be suggested that if used at all, it should refer to *a syncytium that constitutes a whole organism.*

For the sake of consistency it is necessary to classify as syncytia certain temporary arrangements of nuclei and cytoplasm that are not customarily so regarded. In all cases in which nuclei are formed after mitosis before the cytoplasm has completely divided, a short-lived syncytium may be said to exist. In many (perhaps most) animals the male and female pronuclei do not fuse at fertilization; the two sets of chromosomes only come together at the two-cell stage. We have here an example of a short-lived syncytium with two haploid nuclei. As is well known, the process is carried much farther in some copepods, for double nuclei are sometimes seen up to the 32-blastomere stage, and indications of gonomery may persist even later (Häcker, 1892, 1895; Rückert, 1895).

Cells connected by plasmodesmata should not generally be regarded as constituting a syncytium, because the protoplasm is not obviously continuous. Although, as Fol (1896, p. 211) remarks, there is no sharp distinction between cellular tissue and syncytium, yet doubtful cases are rare. It was pointed out by Pringsheim long ago (1860, pp. 229–30) that the hyphae of certain Saprolegniaceae are constricted at intervals, but not divided right across; usually there is only one nucleus between each constriction and the next. Reinhardt (1892, p. 562) described a similar arrangement in *Peziza* and claimed that a part of the streaming cytoplasm passed through a central opening in the transverse wall. If so, the 'cells' clearly constitute a syncytium.

Syncytia appear to have been first noticed in plant tissues by Bauer in the style of *Bletia tankervilliae* (Orchidaceae) in 1802. His drawings were not published till much later; the book evidently appeared in sections that were not separately dated (Bauer, 1830–8). In one of the drawings, here reproduced as fig. 7, the loose tissue of the stigmatic canal after fertilization is shown.

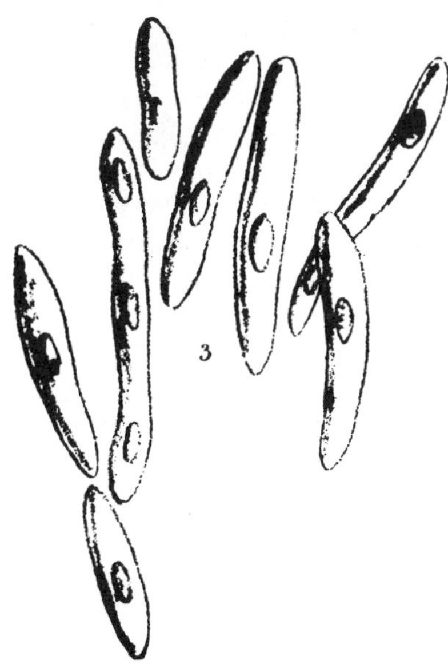

FIG. 7. Bauer's drawing of cells in the stigmatic canal of *Bletia tankervilliae*. One 'cellule' contains two and another three nuclei. The drawing was made in 1802. (Bauer, 1830–8, 1st part ('Fructification'), plate 6, fig. 3.)

Bauer noticed that there were from one to three 'specks' in each of the 'cellules'. It seems just possible that the 'cellules' were sections of pollen-tubes. This drawing by Bauer was known to Brown (1833, p. 711), who mentioned it in the famous paper in which he introduced the word *nucleus*. Brown confirmed Bauer's finding and remarked that it was the only example known to him of more than one nucleus in a cell. Meyen (1837, p. 208) reported that two or three nuclei often occur in 'langgestreckten Zellen'. Unger (1841, fig. 6 on plate V) gave a representation of a cell in the root of *Saccharum*, elongating to form part of a vessel; it is in fact a syncytium containing three nuclei. In their botanical textbook Endlicher and Unger (1843, pp. 22–23) remark, like Meyen, that there are often several nuclei in 'langgestreckten Zellen'. Nägeli (1844, p. 62) mentioned the existence of more than one nucleus, without cellular partitions, in pollen-grains, in the pollen-tube, and in the embryo-sack. The more recent literature of syncytia in the vegetative cells of phanerogams has been reviewed by Beer and Arber (1920).

Syncytia were recorded in the gill-cartilages of *Pelobates* and *Rana* by Schwann (1839, p. 23 and fig. 8 on plate 1). It is clear also that Rathke saw the syncytial stage in the development of the crustacean egg. He expresses himself rather obscurely, but this is only to be expected, as the partial cleavage of a centrolecithal egg had never previously been described. In the passage that follows, which is translated from the Latin, he is referring to the cells seen after cleavage. 'In fact', he writes (1844, p. 8), 'before the cells that I have already mentioned originate, there is formed for each of them, among the structural elements of the yolk [i.e. among the yolk-globules], a particular nucleus, consisting of a vesicle filled with a coagulable liquid. As a result of this, every cell now formed is provided with its own nucleus.' Thus Rathke saw many nuclei in the egg before radiating partitions had appeared to divide it (imperfectly) into blastomeres. Kölliker professed to find syncytia in the embryos of certain invertebrates in which cleavage is in fact total, but stated correctly that more than one nucleus occurs in the giant-cells (polykaryocytes) of bone-marrow and in certain nerve-cells (Kölliker, 1852, pp. 20–21 and fig. 7).

It is interesting to trace the course of events as naturalists were groping their way towards the discovery that a whole organism might consist of a continuous mass of protoplasm containing many nuclei. This truth was first revealed by the investigators of the Mycetozoa. It was known to de Bary (1860) that the plasmodium is formed by the fusion of nucleate cells and that each of the chambers of the sclerotium may contain more than one nucleus, but he did not describe nuclei in the plasmodial stage itself. Schultze, however, in the same year described the plasmodia of *Aethalium septicum* as consisting of 'naked lumps of protoplasm, naturally with the nuclei appertaining to them' (1860, p. 301). It is strange that Schultze should have made this discovery, for he undertook no very detailed study of Mycetozoa, and stranger still that some years later Cienkowski (1863, p. 436), as has been mentioned, and de Bary (1864, p. 108) were still of the opinion that the plasmodium lacks nuclei.

Meanwhile events had been leading towards the discovery of the syncytial nature of certain Heliozoa. Many years before, Kölliker (1849) had seen the nuclei of *Actinosphaerium eichhornii*, and described them as 'kern- und zellenartige' bodies (p. 211); but he did not regard them as nuclei. Stein, too, saw them, and remarked that they had 'das Ansehen von Zellenkernen' (1854, pp. 153–4); but he does not pronounce upon the matter. Haeckel thought the objects were probably nuclei, and suggested that bodies within them might be nucleoli (1862, p. 165). Schultze calls them 'zellenartige Körperchen' in one place and 'Kerne' in another; he makes no definite statement as to what they are (1863, pp. 35–36). It was Wallich who first stated definitely that these bodies in *Actinosphaerium* are nuclei; he uses the word 'nucleus', and defines it (1863, pp. 444, 450; see also his fig. 2 on plate X). Two years later Cienkowski (1865) saw and figured several nuclei in another Helizoon, *Nuclearia delicatula*, and recognized their nature. (It should be remarked that the authors mentioned above called *Actinosphaerium* '*Actinophrys*'.)

In the Radiolaria, also, the numerous nuclei present in certain species were seen long before they were recognized as such. Müller saw them first, in *Acanthometra* (1859, p. 15); he reported the existence of 'many round transparent vesicles' within the central capsule. They were later seen by Haeckel in the Acanthometrida (1862, pp. 141 and 374, and fig. 2 on plate XV), and in the monopylarian *Lithomelissa* (p. 302); he did not know they were nuclei, though he suggested that some Radiolaria might be multinucleate (p. 165). The numerous nuclei of certain Acanthometrida and of *Tridictyopus* (Monopylaria) were for the first time recognized as such by Hertwig (1879, pp. 11, 84), who stained them with carmine. His conclusion was accepted by Haeckel (1887, pp. 32–33).

The syncytial nature of certain other rhizopods was gradually disclosed by the labours of many investigators. Among the thallophytes, on the contrary, most of the important discoveries were made in a short time by one man. Till towards the end of the seventies many of the lower plants were universally regarded as non-nucleate (see, e.g., Haeckel, 1874, p. 409; Sachs, 1874, p. 273; Strasburger, 1876, pp. 86–88; Haeckel, 1878, p. 53). Sporadic discoveries of syncytia had indeed been made. Pringsheim, for instance, as we have seen (p. 178), had described this condition in the Saprolegniaceae, and de Bary (1862, p. 14) had seen stages in the development of the ascus of *Peziza*, with two, four, and eight nuclei not separated by cell-walls. The existence of whole groups of syncytial plants was, however, unsuspected. The plants were known, but their nuclei were not; for though they had been seen in some cases (e.g. in *Cladophora*), they were not regarded as nuclei, simply because there were many of them in each 'cell' (Strasburger, 1876, pp. 86–88 and 324). They were first recognized as such by Schmitz, who revolutionized knowledge of the lower plants by his discoveries. The latter were all published in the journals of local natural history societies. Schmitz used simple methods, staining the nuclei sometimes with alcoholic iodine solution, sometimes with a mixture of haematoxylin, alum, and glycerine after fixation in alcohol or osmium tetroxide solution. He began by studying seaweeds in the Gulf of Athens in 1878, and at Naples in the following year. He reported his first results verbally on 30 November 1878, but there was delay in the printing of this particular paper (Schmitz, 1880*a*). Meanwhile he had published others. He first announced his discovery of nuclei in *Valonia* and related forms, and showed that each of the apparent 'cells' is a syncytium; he founded the group Siphonocladiaceae for these plants, which had up till then been variously classified (1879*a*, 1880*a*). He noted the formation of uninucleate zoospores. Turning next to the Siphonales, he showed the existence of numerous nuclei in the continuous cytoplasm of these non-septate forms (1879*b*). In the same paper he showed the syncytial nature of several Phycomycetes and of the internodes of *Chara*. He also noticed syncytia in parenchyma cells of certain higher plants. He continued his work and showed that among the Rhodophyceae, the species differ in their nuclear arrangements; in one species the 'cell' may be multinucleate, in another closely related form it may be uninucleate (1880*b*).

The wholly syncytial plants and animals show that the cell, as a morphological unit, is not a necessary component of organisms. It is possible, however, to exaggerate the importance of this fact. No organism reaches a high degree of complexity without adopting cellular structure. Some of the Siphonales show a limited degree of resemblance to higher plants in external form, and there is thus a suggestion of a much greater degree of differentiation of parts than in fact exists. Pressed specimens, too, tend to look more like higher plants than do these lowly organisms in their natural form. It is doubtful whether *Caulerpa* and its allies are in fact the most highly differentiated syncytial organisms. Some of the Ciliophora appear more complex. This most aberrant group, however, is excluded from consideration in the present paper, as it will be discussed under Proposition VI. Reasons for regarding the ciliates and their allies as syncytia will be given there. (See also Baker, 1948, *b* and *c*.)

At a meeting of a scientific society in Würzburg on 23 November 1878 Sachs demonstrated a series of Siphonales and remarked that these, as well as the Mucorineae, had up till then been regarded as 'einzellige'; he considered that they should rather be called 'nicht cellulare' (Sachs, 1879). It is to be noted that this demonstration took place seven days before Schmitz began to make known his discoveries on syncytial plants. Like everyone else, Sachs considered that the plants he called non-cellular were devoid of nuclei. When their nuclei were discovered, the name stuck to them; and indeed it is not inappropriate. It is unfortunate that the name was also applied by some writers to certain uninucleate protists, which evidently correspond in their structure to single cells. This matter will be discussed with Proposition VI; I have already commented upon it elsewhere (Baker, 1948, *b* and *c*).

Sachs reverted to the subject of non-cellular plants many years later in two important papers (1892 and 1895). He considered that the accepted terminology of cytology was misleading and should be changed. For him, the cell was the cell-wall, or sometimes the cell-wall with the contents of the cell (1892, p. 62; see also p. 166 of the present paper). He felt that a new word was required. 'Under the name of an Energid', he wrote (1892, p. 57), 'I think to myself of a single cell-nucleus with the protoplasm controlled (beherrschten) by it.' He chose the word *energid* to indicate that the vital activities reside in the nucleus and cytoplasm; he did not intend 'energy' to be understood here in its physical sense (1895, p. 410). He regarded the energid as a morphological as well as a physiological unit. It might produce a cell-wall or other secreted objects, or it might not. In most cases each cell is inhabited by one energid, but the Siphonales were obviously peculiar in this respect. Sachs had changed his mind; he now called them one-celled plants, but remarked that the cell was produced by numerous energids. The difference between such forms as the Siphonales on the one hand and cellular plants on the other, in Sachs's terminology, was that the neighbouring energids of the former were not sharply marked off from one another (1892, p. 62; 1895, p. 425).

Sachs's new word was never widely accepted. Biologists might perhaps do well to reconsider their tacit rejection of it. His expression 'beherrschten' has gained rather than lost in significance since his time. A difficulty is that in syncytia we have generally no means of delimiting the nuclear zones of control, which must be constantly shifting in cases where the cytoplasm is in motion. Again, it is difficult to be certain that a particular part of the cytoplasm is 'controlled' by only one nucleus. Occasionally, however, the zones of control announce very clearly that they exist. A good example is provided by the syncytial zoospore of *Vaucheria*, in which two flagella are related to each nucleus.

Polyploidy

Boveri (1905) called the nucleus of the spermatozoon or egg a *Hemikaryon*; the fusion-nucleus of a zygote an *Amphikaryon*; and a nucleus in which the number of chromosomes had doubled without nuclear division a *Diplokaryon*. Strasburger was using the words *diploid* and *haploid* in 1907 (pp. 490, 529) and *tetraploid*, *oktoploid*, and *polyploid* in 1910 (pp. 422, 444), but it seems doubtful whether we have any authoritative statement of their meanings. A quotation from a well-known and excellent textbook will exemplify the doubt. Its author writes: 'The lowest diploid number found in any organism is 2, which occurs in the Roundworm, *Ascaris megalocephala* var. *univalens* (this species also has a tetraploid variety, *bivalens* with 4 chromosomes in the diploid set).' Thus in one sentence we are told that four is both the diploid and the tetraploid number of chromosomes in the variety *bivalens*. Some authors use the expression diploid as synonymous with the somatic number in sexually produced organisms, and haploid as synonymous with the gametic number, whether or not there happen to be only two sets of different chromosomes in the cells called diploid and one in those called haploid. I shall not follow this usage, but shall employ the word haploid to refer to a single set of different chromosomes and diploid to refer to two such sets, and shall use triploid, polyploid, &c., in conformity with this system. (Thus a gamete-nucleus is characteristically haploid, but may be diploid.)

The haploid protoplast, or *haplocyte* as I shall call it, is a better example of a unit than a *diplocyte*, with its two sets, just as a box containing the playing-cards of a single suit is more perfectly unitary than one containing two suits or the tetraploid pack; but since the great majority of organisms arise, directly or indirectly, from the fusion of two haplocytes, the most usual morphological unit in both plants and animals is the diplocyte. The degree of duplicity exhibited by this cellular unit is an expression of one of the most fundamental facts of biology. The *Uebereinstimmung* postulated by Schwann was to some extent upset by the discovery that there are typically both haplocytes and diplocytes in organisms; but much more serious from the point of view of the cell as a morphological unit is the fact that polyploid nuclei also exist. The history of this discovery must now be briefly traced.

Guignard (1884, p. 27) suggested that certain plant-cells may contain about twice the usual number of 'bâtonnets chromatiques', but he was dealing with a normally haploid structure, and the discovery of polyploidy must be ascribed to Boveri (1887). The latter showed that there are two varieties of *Ascaris megalocephala*, which he called Typus Carnoy and Typus van Beneden, after earlier students of the chromosomes of this nematode. He showed that in Typus Carnoy there are two chromatic elements (chromosomes) in the ripe egg, in Typus van Beneden only one. Thus the somatic cells of the former variety were tetraploid. This case is not quite so simple as it appeared to be, for, as is well known, the chromosomes break into fragments in cells other than those of the germ-track; and when they have done so, the number in Typus van Beneden appears not to be exactly half that in Typus Carnoy (Walton, 1924). It seems allowable, however, in the present state of knowledge, to regard Typus Carnoy as tetraploid before fragmentation.

Boveri later (1903, *a* and *b*, 1905) shook the eggs of the sea-urchin *Strongylocentrotus* immediately after fertilization and by this means suppressed the first cleavage, while the chromosomes divided; he thus obtained tetraploid larvae experimentally.

A discovery in plants similar to Boveri's in *Ascaris* was made by Rosenberg (1903), who showed that *Drosera rotundifolia* has 20 chromosomes in its somatic cells, while *D. longifolia* has 40. Strasburger (1905) counted the number of chromosomes in the pollen mother-cells of *Alchemilla arvensis* and *A. speciosa*, and found that the latter had twice as many as the former. These were the first indications of what is now known to be the very widespread occurrence of polyploidy among higher plants. A complication is introduced by certain species which seem to have become secondarily diploid, with double the usual diploid number of chromosomes, by differentiation of the four sets into two.

Difficulties in the sex-determining mechanism prevent most dioecious animals from doubling their chromosome numbers throughout their tissues, but there are indications that certain hermaphrodites are polyploid (see White, 1940). There is no particular barrier against chromosome replication in somatic cells, and it is not unusual for some of these to become polyploid. The classical example of mosaic polyploidy is provided by the honey-bee. Petrunkevitch found long ago (1901, pp. 587–8) that there are only 16 chromosomes in the first division of the nucleus of the drone-egg, but about 64 in cells of the blastoderm of the later embryo. In a celebrated paper Meves showed that while the diploid number (counted in the oogonia of the queen) is 32 and the haploid 16, more than 60 chromosomes are present in the follicle-cells of the testis (1907, pp. 471–2). Nowadays we have simpler methods of detecting polyploidy, in particular cases, than laborious chromosome-counts. We may measure nuclear volumes (Jacobj, 1925), or count either the nucleoli (especially where there is one per haploid set in early prophase (de Moll, 1923, 1928)), or the heterochromatic X-chromosomes (Geitler, 1937), or heterochromatic satellites (Berger, 1941). The most extreme instance of mosaic polyploidy seems to be provided by the pond-skater, *Gerris lateralis*, in which the degree of replication reaches 1024-ploidy, or even farther, in the salivary glands (Geitter, 1938).

There is evident similarity between multinucleate conditions and polyploidy. There is general correspondence between a single mass of cytoplasm containing two diploid nuclei and another containing one tetrapoid nucleus: both may be called tetraplocytes. Particular tissues tend to provide examples of both the binucleate and the uninucleate tetraploid states. Thus in the roots of *Pisum sativum* treated in life with chloral hydrate, Strasburger (1907, pp. 484–5) found both mitoses with tetraploid chromosome-numbers and binucleate cells. In the tapetal layer of the anthers of certain plants, some cells show polyploid chromosome-numbers at division, while others are bi- or multinucleate (see especially Witkus's study of *Spinacia* (1945)). In certain parts of young seedlings of *Allium cepa*, again, a number of tetraploid cells are formed in certain regions; in the same sites binucleate cells are common (Berger and Witkus, 1946). In the mammalian liver there are polyploid cells of various degrees of chromosome-replication; there are also bi- and multinucleate cells (Wilson and Leduc, 1948). Fell and Hughes (1949, p. 366) have shown by the study of living tissue-culture cells of the mouse that polyploid nuclei may arise by mitosis of binucleate cells, a single spindle being formed for the chromosomes of the two nuclei; fusion of diploid nuclei and endomitosis are other mechanisms by which the same end is achieved (Berger, 1937; Geitler, 1939; Wilson and Leduc, 1948).

The 'polyenergid' nuclei of certain Protozoa will be mentioned later under the heading of Proposition VI. I have already discussed them elsewhere (Baker, 1948*b*).

The existence of polyploidy undoubtedly constitutes an exception to the general rule of the 'Uebereinstimmung' of all cells. One polyploid cell cannot be regarded as homologous with one diploid cell. Boveri (1903*a*), for instance, found that the protoplasts of tetraploid larvae of *Strongylocentrotus* were much larger and fewer than those of the diploid form; in the case of the mesenchyme he found that there was about half the normal number. Thus one tetraploid corresponds with two diploid protoplasts. Polyploidy, however, is clearly a secondary condition. Diplocytes and haplocytes are the characteristic primitive morphological units of plants and animals, and are still retained as the elementary components of most organisms. Those organisms that show mosaic polyploidy have haploid germ-cells and are everywhere diploid in the early embryonic stage of the sexually produced form.

The Indivisibility of Cells into Smaller Homologous Units

Hirsch remarks (1942) that the body of an organism consists of a series of 'partial systems', each of which (till protons and electrons are reached) is built of partial systems of a lower order. Thus the body is made up of organs which are divisible into tissue-units and these into cells; the latter contain *Mikronen* (mitochondria, Golgi-bodies (lipochondria), various granules and vacuoles, muscle-fibrils, chromosomes, nucleoli), and these are composed of sub-microscopic *Submikronen*, themselves made up of large molecules; and so on.

If an object is composed of parts, all of which are divisible into smaller parts that show what Schwann called 'Uebereinstimmung' with one another, then these smaller parts are clearly the true units of construction. It is important to stress the fact that the Mikronen are not homologous parts: a muscle-fibril, for instance, does not correspond, in any predicable way, to a nucleolus. Further, the Mikronen taken together do not constitute the cell, which consists largely of ground-cytoplasm and nuclear sap. There is no intention here to criticize Hirsch's analysis adversely, but only to point out that it in no way invalidates the cell-theory. *The cell is not composed of any lesser homologous units*, other than those minute particles that compose all matter, and to these the idea of homology does not properly apply. As Hanstein remarked long ago: 'In the last resort the *protoplast, not* the *molecule* or the *micelle*, is the organic individual' (1880, p. 295). With the reservations that have already been noted, this is true.

Comment

Adequate critiques have already been given in this paper of the various facets of the cell-theory that are included under the head of the second part of the second Proposition. It remains to make one general comment. Whenever a student wants to 'understand' a complex histological object that is unfamiliar to him (a Pacinian corpuscle will serve as an example) or a research-worker to grasp the minute structure of a previously unknown organ, he proceeds first of all to try to determine where the boundaries of the protoplasts are—what is cellular and what is intercellular. In other words, he tries to interpret what he sees in terms of the part of the cell-theory that is summarized in Proposition II, in the formulation here adopted. That is the measure of the homage he pays (often unwittingly) to the founders of the cell-theory.

Acknowledgements

Professor A. C. Hardy, F.R.S., has given all the necessary facilities for carrying out this work. I am indebted to Dr. C. F. A. Pantin, F.R.S., for valuable criticism of the first draft of this paper. Among members of the staffs of several libraries who have helped me, I must particularly mention Miss E. S. Ford and Miss R. Guiney. Miss L. M. Newton kindly showed me part of the collection of Siphonales in the British Museum (Natural History). Mrs. J. A. Spokes and Miss B. M. Jordan have given careful clerical assistance.

REFERENCES

ANDREWS, E. A., 1899. Zool. Bull., **2**, 1.

ANDREWS, G. F., 1897. Journ. Morph., **12**, 367.

BAKER, J. R., 1948*a*. Quart. J. micr. Sci., **89**, 103.

—— 1948*b*. Nature, **161**, 548.

—— 1948*c*. Ibid., **161**, 587.

—— 1949. Quart. J. micr. Sci., **90**, 87.

BARY, A. DE, 1860. Zeit. wiss. Zool., **10**, 88.

—— 1863. *Über die Fruchtentwicklung der Ascomyceten. Eine pflanzenphysiologische Untersuchung.* Leipsig (Engelmann).

—— 1864. *Die Mycetozoen (Schleimpilze). Ein Beitrag zur Kenntniss der niedersten Organismen.* Leipsig (Engelmann).

BAUER, F., 1830–8. *Illustrations of orchidaceous plants.* London (Ridgway).

BEER, R., and ARBER, A., 1920. Journ. roy. micr. Soc. (no vol. number), 23.

BERGER, C. A., 1937. Amer. Nat., **71**, 187.

—— 1941. Ibid., **75**, 93.

—— and WITKUS, E. R., 1946. Amer. J. Bot., **33**, 785.

BIZZOZERO, G., 1876. Untersuch. Naturlehre (Moleschott) **11**, 30.

BOEKE, J., 1940. *Problems of nervous anatomy.* Oxford (Univ. Press).

BOURNE, G. C., 1895. Quart. J. micr. Sci., **38**, 137.

BOVERI, T., 1887. *Zellen-Studien.* Heft 1. *Die Bildung der Richtungskörper bei Ascaris megalocephala und Ascaris lumbricoides.* Jena (Fischer).

—— 1903*a*. Verh. phys.-med. Ges. Würzburg, **35**, 67.

—— 1903*b*. Sitz.-Ber. phys.-med. Ges. Würzburg (no vol. number), 12.

—— 1905. *Zellen-Studien.* Heft 5. *Ueber die Abhängigkeit der Kerngrösse und Zellenzahl der Seeigel-Larven von der Chromosomenzahl der Ausgangszellen.* Jena (Fischer).

BOZLER, E., 1927. Zeit. Zellforsch. mikr. Anat., **5**, 244.

BRISSEAU-MIRBEL, —, 1808. *Exposition et défense de ma théorie de l'organization végétale.* La Haye (van Cleef).

—— 1835. See Mirbel.

BROWN, R., 1833. Trans. Linn. Soc., **16**, 685.

BRÜCKE, E., 1862. Sitz. Kais. Akad. Wiss. Wien, **44**, 2, 381.

BÜTSCHLI, O., 1883. H. G. Bronn's *Klassen und Ordnungen des Thier-Reichs.* Erster Band, II Abtheilung: *Mastigophora.* Leipsig und Heidelberg (Winter).

CIENKOWSKI, L., 1863*a*. Jahrb. wiss. Bot., **3**, 325.

—— 1863*b*. Ibid., **3**, 400.

—— 1865. Arch. mikr. Anat., **1**, 203.

COHN, F., 1875. Beitr. Biol. Pfl., **1**, 3, 93.

DANIELLI, J. F., 1951. Chapter on *The cell surface and cell physiology*, in *Cytology and cell physiology*, edited by G. Bourne. Oxford (Clarendon Press).

DAVIS, B. M., 1905. Amer. Nat., **39**, 217.

DAVSON, H., and DANIELLI, J. F., 1943. *The permeability of natural membranes.* Cambridge (Univ. Press).

DELAGE, Y., and HÉROUARD, E., 1896. *Traité de Zoologie concrète*, vol. 1. Paris (Schleicher).

DUHAMEL DU MONCEAU, —, 1758. *La physique des arbres; où il est traité de l'anatomie des plantes et de l'économie végétale.* 2 vols. Paris (Guerin and Delatour).

DUTROCHET, M. H., 1837. *Mémoires pour servir à l'histoire anatomique et physiologique des végétaux et des animaux.* Paris (Baillière).

EBNER, V. v., 1896. Sitz. Akad. Wiss. Wien, **105**, 3, 123.

ELSBERG, L., 1883. Quart. J. micr. Sci., **23**, 87.

ENDLICHER, S., and UNGER, F., 1843. *Grundzüge der Botanik.* Wien (Gerold).

ESAU, K., 1939. Bot. Rev., **5**, 373.

FELL, H. B., and HUGHES, A. F., 1949. Quart. J. micr. Sci., **90**, 355.

FLEMMING, W., 1896. Anat. Hefte, 2. Abt., **6**, 184.

FOL, H., 1896. *Lehrbuch der vergleichenden mikroskopischen Anatomie.* Leipsig (Engelmann).

FOREL, A., 1887. Arch. f. Psychiat. und Nervenkrank., **18**, 162.

FROMMAN, C., 1879. Sitz. Jen. Ges. Med. Naturw. (no vol. number), 51.

FULIŃSKI, B., 1916. Zool. Anz., **47**, 381.

GARDINER, W., 1884. Arb. bot. Inst. Würzburg, **3**, 1, 52.

GEGENBAUR, C., 1874. *Grundriss der vergleichenden Anatomie.* Leipsig (Engelmann).
GEITLER, L., 1937. Zeit. Zellforsch. mikr. Anat., **26**, 641.
—— 1938. Biol. Zentralbl., **58**, 152.
—— 1939. Chromosoma, **1**, 1.
GOEBEL, K. VON, 1926. *Wilhelm Hofmeister: the work and life of a nineteenth-century botanist.* London (Ray Soc.).
GOLGI, C., 1883. Arch. ital. de Biol., **3**, 285.
GOROSCHANKIN, J., 1883. Bot. Zeit., **41**, col. 825.
GREW, N., 1672. *The anatomy of vegetables begun.* London (Hickman).
—— 1682. *The anatomy of plants.* (Published by the author; place not stated.)
GROSS, J., 1901. Zeit. wiss. Zool., **69**, 139.
—— 1903. Zool. Jahrb. Abt. Anat., **18**, 71.
HÄCKER, V., 1892. Zool. Jahrb. Abt. Anat., **5**, 211.
—— 1895. Arch. mikr. Anat., **45**, 339.
—— 1895. Ibid., **46**, 579.
HAECKEL, E., 1862. *Die Radiolarien.* Berlin (Reimer).
—— 1866. *Generelle Morphologie der Organismen.* Berlin (Reimer).
—— 1872. *Biologie der Kalkschwämme (Calcospongien oder Grantien).* Berlin (Reimer).
—— 1874. *Natürliche Schöpfungsgeschichte,* 5th edit. Berlin (Reimer).
—— 1878. *Das Protistenreich.* Leipsig (Günther).
—— 1887. *Grundriss einer allgemeinen Naturgeschichte der Radiolarien.* Berlin (Reimer).
—— 1894. *Systematische Phylogenie.* Berlin (Reimer).
—— 1898. *Natürliche Schöpfungsgeschichte,* 9th edit. Berlin (Reimer).
HAMMAR, J. A., 1896. Arch. mikr. Anat., **47**, 14.
—— 1897. Ibid., **49**, 92.
HANSTEIN, J. V., 1880. *Das Protoplasma als Träger der pflanzlichen und thierischen Lebensverrichtungen. Für Laien und Sachgenossen dargestellt.* From *Sammlung von Vorträgen für das deutsche Volk,* edited by W. Frommel and F. Pfaff, p. 125. Heidelberg (Winter).
HARTIG, T., 1854. Bot. Zeit., **12**, col. 51.
HEITZMANN, C., 1873. Sitz. Akad. Wiss. Wien, **67**, 3, 141.
—— 1883. *Mikroskopische Morphologie des Thierkörpers im gesunden und kranken Zustande.* Wien (Braumüller).
HENLE, J., 1841. *Allgemeine Anatomie. Lehre von den Mischungs- und Formbestandtheilen des menschlichen Körpers.* Leipsig (Voss).
HERTWIG, R., 1879. *Der Organismus der Radiolarien.* Jena (Fischer).
HIRSCH, G. C., 1942. *Der Bau des Tierkörpers*: vol. 5, part 1, of *Handbuch der Biologie,* edited by L. von Bertalanffy. Potsdam (Athenaion).
HIS, W., 1886. Abh. math.-phys. Classe Kön. Sächs. Ges. Wiss., **13**, 477.
—— 1889. Ibid., **15**, 311.
IHERING, H. VON, 1877. Zeit. wiss. Zool., **29**, 1.
JACOBJ, W., 1925. Arch. f. Entw., **106**, 124.
JANET, C., 1912. *Le Volvox.* Limoges (Ducourtieux et Gout).
KIENITZ-GERLOFF, F., 1891*a*. Bot. Zeit., **49**, col. 17.
—— 1891*b*. Ibid., **49**, col. 33.
KÖLLIKER, A., 1852. *Handbuch der Gewebelehre des Menschen. Für Aerzte und Studirende.* Leipsig (Engelmann).
KOROTNEFF, A., 1876. Arch. Zool. exp. gén., **5**, 203.
KORSCHELT, E., 1886. Zeit. wiss. Zool., **43**, 537.
—— 1936. *Vergleichende Entwicklungsgeschichte der Tiere.* Vol. 1. Jena (Fischer).
KUHLA, F., 1900. Bot. Zeit., **58**, 29.
KULTSCHIZNY (*sic,* should be Kultschitzky), N., 1888. Biol. Centralbl., **7**, 572.
LEYDIG, F., 1857. *Lehrbuch der Histologie des Menschen und der Thiere.* Frankfurt (Meidinger).
LINK, [D.] H. F., 1807. *Grundlehren der Anatomie und Physiologie der Pflanzen.* Göttingen (Danckwerts).
—— 1809. *Nachträge zu den Grundlehren der Anatomie und Physiologie der Pflanzen.* 1st Heft. Göttingen (Danckwerts).
—— 1812. Ibid., 2nd Heft. Göttingen (Danckwerts).
MALPIGHI, M., 1675. *Anatome plantarum. Cui subjungitur appendix, iteratas & auctas ejusdem authoris de ovo incubato observationes continens.* London (Roy. Soc.).

MEISSNER, G., 1856. Zeit. wiss. Zool., **7**, 1.
MESTSCHERSKAYA, K., 1931. Zeit. Zellforsch. mikr. Anat., **13**, 109.
MEVES, F., 1907. Arch. mikr. Anat., **70**, 414.
MEYEN, F. J. F., 1837. *Neues system der Pflanzen-Physiologie.* Berlin (Haude und Spenersche Buchhandlung).
MEYER, A., 1895. Bot. Centralbl., **63**, 225.
—— 1896. Bot. Zeit., **54**, 187.
MIRBEL, —, 1835. Mém. Acad. Roy. Sci. Inst. de France, **13**, 337.
MOLDENHAWER, J. J. P., 1812. *Beyträge zur Anatomie der Pflanzen.* Kiel (Wäser).
MOLL, W. E. DE, 1923. Genetica, **5**, 225.
—— 1928. La Cellule, **38**, 5.
MÜLLER, J., 1859. Phys. Abh. Kön. Akad. Wiss. Berlin (no vol. number), 1.
NÄGELI, C., 1844. Zeit. wiss. Bot., **1**, 1, 34.
NICOL, J. A. C., 1948. Quart. J. micr. Sci., **89**, 1.
NONIDEZ, J., 1944. Biol. Rev., **19**, 30.
OKEN, L., 1805. *Die Zeugung.* Bamberg u. Wirzburg (Goebhardt).
OVERTON, E., 1895. Vierteljahrsschr. naturf. Ges. Zürich, **40**, 159.
—— 1899. Ibid., **44**, 88.
—— 1900. Jahrb. wiss. Bot., **34**, 669.
PANTIN, C. F. A., 1935*a*. Journ. exp. Biol., **12**, 119.
—— 1935*b*. Ibid., **12**, 139.
PETRUNKEWITSCH, A., 1901. Zool. Jahrb. Abt. Anat., **14**, 575.
PRINGSHEIM, N., 1854. Reprinted in *Gesammelte Abhandlungen von N. Pringsheim*, 1896, vol. 3, p. 33.
—— 1860. Jahrb. wiss. Bot., **2**, 205.
RAMÓN Y CAJAL, S., 1934. Trav. Lab. Rech. biol. Univ. Madrid, **29**, 1.
RATHKE, H., 1844. *De animalium crustaceorum generatione.* Regiomonti (Dalkowski).
REICHERT, C., 1841. Arch. Anat. Physiol. wiss. Med. (no vol. number), 523.
REINHARDT, M. O., 1892. Jahrb. wiss. Bot., **23**, 479.
REMAK, R., 1851. Tagsber. Fort. Heilk. (Froriep), Abt. Anat. Physiol., **1**, 316.
—— 1855. *Untersuchungen über die Entwickelung der Wirbelthiere.* Berlin (Reimer).
—— 1862. Arch. f. Anat. Physiol. wiss. Med. (no vol. number), 230.
ROSENBERG, O., 1903. Ber. deut. bot. Ges., **21**, 110.
RÜCKERT, J., 1895. Arch. mikr. Anat., **45**, 339.
RUSSOW, E., 1881. Sitz. Naturf. Ges. Univ. Dorpat, **6**, 63.
—— 1884. Ibid., **6**, 562.
SACHS, J., 1863. Flora, **21**, 65.
—— 1874. *Lehrbuch der Botanik nach dem gegenwärtigen Stand der Wissenschaft.* Leipsig (Engelmann).
—— 1879. Verk. physikal. med. Ges. Würzburg, **13**, p. xliv.
—— 1892. Flora, **75**, 57.
—— 1895. Ibid., **81**, 405.
SALMON, J., 1946. *Différentiation des tubes criblés chez les angiospermes. Recherches cytologiques.* (Published by the author; place not stated.)
SCHÄFER, E. A., 1878. Phil. Trans., **169**, 563.
SCHMITZ, —, 1879*a*. Sitz. niederrhein. Ges. Bonn (no vol. number), 142.
—— 1879*b*. Ibid., 345.
—— 1880*a*. Ber. Sitz. Naturf. Ges. Halle, **14**, 17.
—— 1880*b*. Sitz. niederrhein. Ges. Bonn (no vol. number), 123.
SCHRÖN, O., 1865. Untersuch. Naturlehre (Moleschott), **9**, 93.
SCHULTZE, M., 1860. Arch. f. Naturges., **26**, 287.
—— 1861. Arch. Anat. Physiol. wiss. Med. (no vol. number), 1.
—— 1864*a*. Centralbl. med. Wiss. (no vol. number), 177.
—— 1864*b*. Arch. path. Anat., **30**, 260.
SCHWANN, T., 1839. *Mikroskopische Untersuchungen.* Berlin (Sander'schen Buchhandlung).
SEDGWICK, A., 1885. Quart. J. micr. Sci., **25**, 449.
—— 1894. Ibid., **37**, 87.
—— 1895. Ibid., **38**, 331.
SIEBOLD, C., 1843. Arch. f. Naturges., **9**, 2, 300.
SMITH, I., 1904. Bot. Gazette, **37**, 346.

STEIN, F., 1854. *Die Infusionsthiere auf ihre Entwickelungsgeschichte untersucht.* Leipzig (Engelmann).
STRASBURGER, E., 1876. *Über Zellbildung und Zelltheilung.* Jena (Dabis).
—— 1901. Jahrb. wiss. Bot., **36**, 493.
—— 1905. Ibid., **41**, 88.
—— 1907. Ibid., **44**, 482.
—— 1910. Flora, **100**, 398.
TANGL, E., 1879–81. Jahrb. wiss. Bot., **12**, 170.
TREVIRANUS, G. R., 1805. *Biologie, oder Philosophie der lebenden Natur für Naturforscher und Aerzte.* Vol. 3. Göttingen (Röwer).
TREVIRANUS, L. C., 1811. *Beyträge zur Pflanzenphysiologie.* Göttingen (Dieterich).
UNGER, [F.], 1841. Linnaea, **50**, 385.
VEJDOVSKY, F., 1888. Zeit. wiss. Zool., **46**, 188.
VRIES, H. DE, 1884. Jahrb. wiss. Bot., **14**, 427.
—— 1885. Ibid., **16**, 465.
WALLICH, G. C., 1863. Ann. Mag. nat. Hist., **11**, 434.
WEBER, C. O., 1858. Arch. path. Anat. (Virchow), **15**, 465.
WEISS, P., 1940. Amer. Nat., **74**, 34.
WHITE, M. J. O., 1940. Nature, **146**, 132.
WILSON, J. W., and LEDUC, E. H., 1948. Amer. J. Anat., **82**, 353.
WITKUS, E. R., 1945. Amer. J. Bot., **32**, 326.
YOUNG, J. Z., 1939. Phil. Trans. B., **229**, 465.

The Cell-theory: a Restatement, History, and Critique

Part IV. The Multiplication of Cells

By JOHN R. BAKER

(From the Department of Zoology and Comparative Anatomy, Oxford)

Summary

In the first half of the nineteenth century it was commonly supposed that new cells arose either *exogenously*, outside pre-existing cells, or *endogenously*, from small rudiments that appeared within pre-existing cells and gradually grew larger. The theory of exogeny had been founded by Wolff (1759), and was supported especially by Link (1807), Schwann (1839), and Vogt (1842). The theory of endogeny, which had been hinted at by various writers in early times, obtained the backing of a very large literature. Its chief advocates were Raspail (1825, &c.), Turpin (1827, &c.), Schleiden (1838), Kölliker (1843–4), and Goodsir (1845).

That cells do not arise exogenously or endogenously, but are produced by the division of pre-existing cells, was at last realized by the convergence of studies made in three separate fields, as follows:

(1) Trembley (1746, &c.), Morren (1830, 1836), Ehrenberg (1830, 1832, 1838), and others noticed how protists multiply.

(2) Dumortier (1832), Mohl (1837), and Meyen (1838) watched the partitioning of the cells of filamentous algae.

(3) Several observers studied the cleavage of eggs and at last revealed that this was a process of cell-division (Prévost and Dumas (1824), von Siebold (1837), Barry (1839), Reichert (1840), Bagge (1841), Bergmann (1841–2)).

Nägeli (1844, 1846) also made an important study of cell-division in all the main groups of plants (except bacteria), but used an unfortunate nomenclature that tended to obscure the true nature of the process.

Remak (1852 and 1855) and Virchow (1852, 1855, 1859) made general statements to the effect that division is the standard method by which cells multiply. The writings of Remak on this subject were much more weighty than those of Virchow.

Contents

[Quarterly Journal of Microscopical Science, Vol. 94, part 4, pp. 407–440, Dec. 1953.]

Introduction

WE are concerned here with the proposition that *cells always arise, directly or indirectly, from pre-existent cells, usually by binary fission*; that is to say, with Proposition III in the formulation of the cell-theory adopted in this series of papers.

With a few exceptions the early cytologists appear not to have been very inquisitive about the way in which cellular structure developed: they were content to describe what they saw at a particular moment in time. About the beginning of the nineteenth century, however, attention began to be focused on the subject of the multiplication of cells. Unfortunately several false theories were promulgated at that time and gained a good deal of acceptance, so that when the truth began to be disclosed towards the middle of the century, by the convergence of unconnected studies, the new discoveries had to contend against firmly established errors. To give a realistic history of the discovery of the actual method by which cells multiply, it is necessary at the outset to present a rather full account of the erroneous views, which were expounded by such distinguished investigators as Wolff, Sprengel, L. C. Treviranus, Raspail, Schleiden, Schwann, and Kölliker. There is a special reason why the exact nature of the errors should be understood. As we have already seen in this series of papers, it happens from time to time that someone alights casually on a particular passage in an old book or journal and attributes a discovery to the author of it, when critical reading and thorough preliminary knowledge would have shown that the writer of the passage actually held entirely mistaken opinions. A careful history of such opinions is necessary if credit is to be restricted to those who really deserve it.

A study of the very extensive literature of the subject reveals that there are three main methods by which cells have been supposed to multiply. These will here be called *exogeny*, *endogeny*, and *division*. By exogeny I mean the origin of new cells outside existing ones; by endogeny, the growth of new cells from small rudiments within an existing cell; and by division, the carving up of an existing cell into two or more smaller ones.

The following classification of the theories of cell-multiplication will be used in the present paper:

Exogeny
- by partitioning
- by vacuolation
- from granules

Endogeny
- with migration from the protoplast
- without migration

Cell-division
- by partitioning

with constriction of the cell-wall
with formation of entirely new cell-walls
in the absence of cell-walls (division of the naked protoplast).

This classification is intended to be as logical, precise, and self-explanatory as possible, but the meaning of its terms must be more fully explained below. These terms were not used by the originators of the several theories or by their adherents. As we shall see, some of the terms used by the early students of the subject are in fact inappropriate, and would confuse the account given here.

It may be remarked at the outset that while exogeny and endogeny are unreal, the various methods of division mentioned in the classification occur in nature.

The present paper deals with the history of the discovery of the methods of cell-multiplication down to the time of general acceptance of the views summarized in the phrase, *Omnis cellula e cellula.* In the next paper in the series it will be necessary to tell the story of the discoveries culminating in the generalization, *Omnis nucleus e nucleo.* The derivation of cell from cell and nucleus from nucleus will lead us back to the cell that originates a new individual. To complete the discussion of Proposition III it will therefore be necessary in the succeeding paper to show how it was discovered that the fertilized ovum is a cell formed by the fusion of two cells.

THE SUPPOSED ORIGIN OF CELLS BY EXOGENY

In Grew's little book, *The anatomy of vegetables begun* (1672), there is an interesting passage bearing on the subject of the origin of cells. It has already been mentioned in Part I of this series of papers (Baker, 1948) that Grew demonstrated the cellular nature of plant-embryos, and he must have realized that the adult plant contains an immensely greater number of cells (or 'Pores', as he often called them). He does not mention this subject specifically, but it seems to have been at the back of his mind when he wrote these words: 'In the *Piths* of many Plants, the greater Pores have some of them lesser ones within them, and some of them are divided with cross Membranes: And betwixt their several sides, have, I think, other smaller Pores visibly interjected' (p. 79). Thus Grew seems to have thought that new cells might originate in various ways. Visible interjection of new pores between the sides of old ones must presumably mean the origin of new cells by exogeny, but unfortunately he gives no details that would enable us to classify this supposed method of cell-multiplication more exactly. These words of Grew constitute the earliest reference to the problem of the multiplication of cells. I have already called attention to them elsewhere (1951, 1952*b*).

Exogeny by partitioning

The supposed origin of cells by exogenous partitioning is illustrated diagrammatically in fig. 1. A space between existing cells enlarges; partitions begin to appear in this space; they become more evident; new cells thus originate, and these enlarge.

The opinion that cells multiply by exogenous partitioning was put forward by Link (1807, p. 31). He later repeated his opinion that new cells originate in this way (1809–12, vol. 1, p. 7). He received little support, however, from subsequent writers, though Mirbel's *développement inter-utriculaire* and *super-utriculaire* (1835, p. 369) may perhaps fall into this category of theories. Mirbel's ideas were confused at the time by his firm belief that the whole of

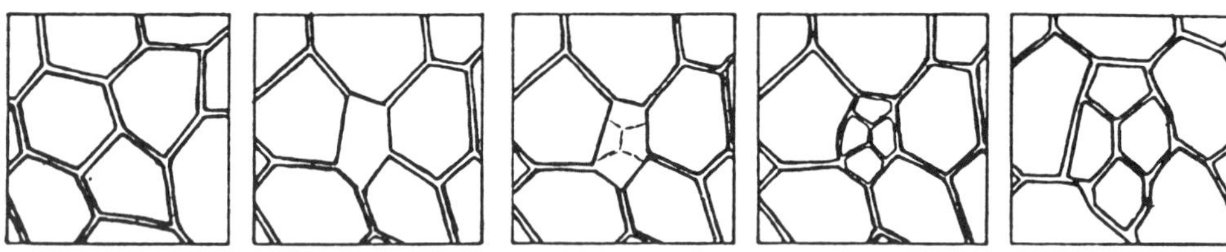

FIG. 1. Diagram of exogeny by partitioning. In this and in all the succeeding diagrams (figs. 2–6), the earliest stage is represented in the square on the left side, and the sequence of events is shown in the remaining squares from left to right across the figure.

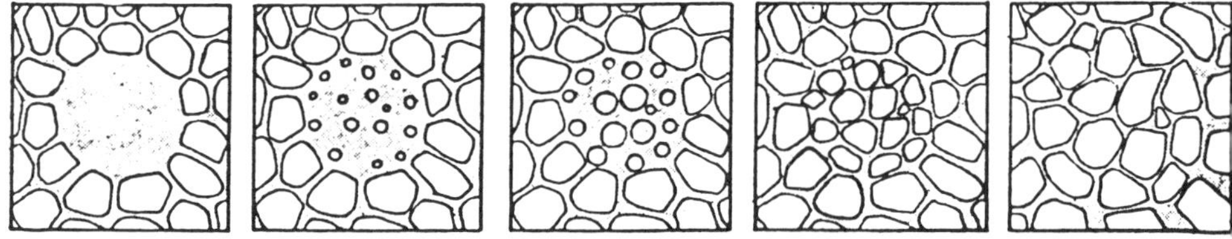

FIG. 2. Diagram of exogeny by vacuolation. The solid intercellular substance is shaded.

the 'membranous tissue' (that is, all the cell-walls) of a plant was perfectly continuous (see Baker, 1952*a*, p. 160), and the meaning of what he wrote on the subject of cell-multiplication is not clear.

Exogeny by vacuolation

This is represented in fig. 2. Between the cells there is a homogeneous, solid or semi-solid substance. In places where there is much of this substance, minute vacuoles sometimes appear in it. These enlarge and transform themselves into new cells resembling the old.

Grew seems to have thought that cells might originate in some such way as this. He remarks (1682, p. 49) that when the sap penetrates into the seed, the liquid internal parts of the latter become coagulated into a solid; a process of 'Fermentation' transforms the coagulum 'into a *Congeries* of *Bladders*: For such is the *Parenchyma* of the whole *Seed*'.

It was Wolff (1759), however, who described exogeny by vacuolation most explicitly. His erroneous views on this subject formed the basis for his theory of epigenesis. He believed that the growing parts of plants were formed of a 'pure, homogeneous, glassy substance' (*pura æquabilis vitrea substantia*, p. 13);

in another place he calls it a 'delicate, solid substance' (*substantia tenera solida*, p. 17). This material permitted the passage of nutritious fluids (p. 17). Minute holes (*punctula*, p. 19), widely separated from one another, were formed in it from blebs (*bullulae*, p. 17) of nutritious fluid; these holes swelled to become cells (*vesiculae*, p. 15; *cellulae*, p. 19). The glassy substance remained as the *interstitia* between them (p. 8). 'Leaves therefore grow for the most part by the interposition of new vesicles between the old, though partly indeed also by the enlargement of the [existing] vesicles' (p. 14).

Wolff did not homologize the *globuli* constituting the blastoderm of the developing hen's egg with the *vesiculae* or *cellulae* of plants. On the contrary, he thought that the material composed of globules, despite its lack of homogeneity, was the counterpart of the glassy interstitial substance of plants; and

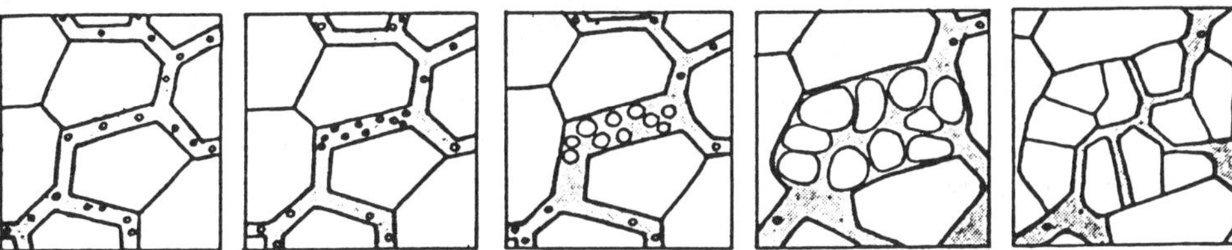

Fig. 3. Diagram of exogeny from granules. The intercellular fluid is shaded.

he thought that the cellular parts of the embryo (the *cellulosa animalis*)—that is, the viscera and vessels—were laid down epigenetically *in* this substance formed of globules (pp. 72, 75).

If one is to hold a balanced view of the history of the epigenesis-preformation controversy, it is necessary to grasp firmly what Wolff's views on the subject of cell-multiplication really were. The opinion, so commonly expressed, that Wolff was essentially right and Bonnet wrong, cannot be substantiated by a study of their writings. I have discussed this matter elsewhere (Baker, 1952*b*, pp. 183–6).

Exogeny from granules

This is represented in fig. 3. Small granules originate in the intercellular fluid; they expand, press upon one another, and become new cells.

It was from a study of the cotyledons in the germination of the seed that Sprengel (1802, pp. 89–90) derived his opinion that new plant cells originated from granules that subsequently enlarged. The granules in the cotyledons were in fact presumably starch-grains. It would appear that in his opinion the cell-forming granules sometimes originated inside and sometimes outside pre-existing cells, but unfortunately he is not explicit on this point. L. C. Treviranus adopted Sprengel's view (1806, pp. 2, 6–10, 14–16). He says that the intercellular spaces of plants contain a fluid that sometimes precipitates fine granules; these grow into *Blasen* or cells. For him, indeed, the purpose of the intercellular fluid (*Saft*) was to produce new cells: it carried the granules wherever new cellular tissue was to be formed. It did not surprise him that the granules that were to become cells were sometimes seen within cells, because he believed that there was free communication between the intercellular fluid and the cavities of the cells, through apertures in the cell-walls.

Rudolphi (1807, p. 35) considered that the intercellular fluid could form new *Bläschen*. He supported Sprengel in general, but gave no particulars.

After a long period of eclipse, the theory of exogeny from granules was reintroduced in the eighteen-thirties and became famous through its promulgation, in a modified form, by Schwann. Valentin (1835, p. 194) first gave a curious account of the origin of the pigment of the chorioid coat of the eye of birds and mammals. Colourless, transparent bodies that he called by the misleading names of *Pigmentkörperchen* and *Pigmentbläschen* appeared first, and the actual globules of pigment subsequently developed in aggregations round each of them. Four years later (1839, p. 133), Valentin announced that the *Pigmentbläschen* were in fact nuclei, and that cells containing pigment were formed round the nuclei after the latter had appeared. This led to a dispute with Schwann (1839, p. 264) about priority.

Schwann, as is well known, considered that new cells originate in a structureless substance which he called the *Cytoblastema* (1839, p. 45) or *Cytoblastem* (p. 112 and elsewhere). This substance, he supposed, sometimes existed within pre-existing cells, but in animals it was usually extracellular (pp. 203–4). Its consistency differed in different cases. It was often fluid, but might also be solid: the matrix of cartilage was an example of it. Schwann's general scheme of cell-formation was as follows (1839, pp. 207–12). The first object to appear in the previously homogenous *Cytoblastem* was the nucleolus. A clump of granules next appeared round this; these then resolved themselves into a pellucid nucleus with a clear boundary, which sometimes took the form of a distinct membrane. The nucleus grew. When it had reached a certain size, a substance derived from the *Cytoblastem* was deposited on it in the form of a layer. Either the whole of this layer, or the outer part of it only, was the future cell-wall (*Membran*). The nucleus adhered to the cell-wall in one place, but elsewhere a fluid appeared between the two and separated them; this fluid, the *Zelleninhalt*, increased in volume. The typical nucleated cell was thus produced. The nucleus in most cases was eventually absorbed and disappeared.

In forming his opinions about the origin of cells, Schwann was undoubtedly much influenced by Schleiden, though he placed the *Cytoblastem* of animal cells, as a general rule, outside pre-existing cells, while Schleiden regarded cell-formation in plants as endogenous (see below, p. 416).

The exogenous origin of cells in a *Cytoblasteme*, as he called it, was reiterated by Vogt in his book on the development of the obstetric toad, *Alytes* (1842). Vogt distinguished between *Cytoblasteme primäre*, or intercellular material that had never formed part of a cell, and *Cytoblasteme secondäre*, formed of material that had previously composed cells and had subsequently become structureless (p. 125). He held that cell-formation started in *Alytes* when cleavage was finished (pp. 9–10, 25). In the *Cytoblasteme* (whether primary or secondary) a nucleus originated, and round this a cell (pp. 117–19); sometimes the cell originated first (pp. 119–20). Vogt does not describe the details of the process, but the nucleolus was not the first object to appear (p. 118).

It is a remarkable fact that so late as 1849, Virchow instituted a comparison between the origin of a crystal from its mother-liquor, and a cell from the *Blastem*. 'Both the mother-liquor and the *Blastem* are amorphous substances, from which bodies of definite shape arise by aggregation of atoms' (1849, pp. 8–9). The difference lay in the substance of the crystal being already present in the liquor, while chemical change was necessary for the differentiation of the cell.

The Supposed Origin of Cells by Endogeny

Endogeny with migration from the protoplast

Fig. 4 represents the origin of a new cell in a filamentous alga by this hypothetical method. In such a form as this there are no intercellular spaces,

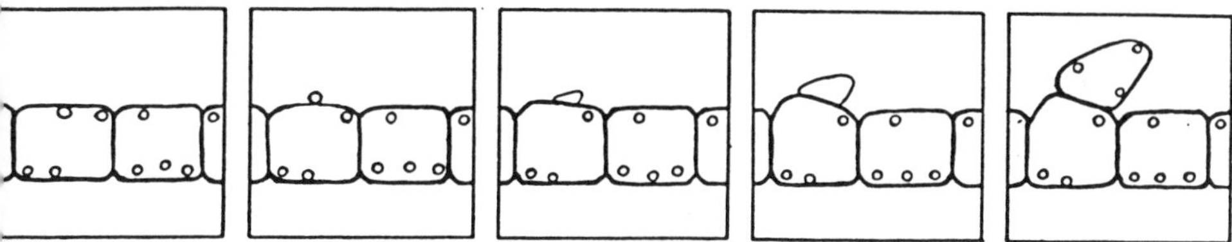

Fig. 4. Diagram of endogeny with migration from the protoplast.

and exogeny is therefore scarcely possible. The cells contain granules. These have the property of being able to migrate through the cell-wall and grow into new cells; in these, new granules appear endogenously, which are capable of repeating the process.

L. C. Treviranus supposed that this method of cell-multiplication occurred in certain algae. He considered that the new tubes (*Schläuchen*) of the water-net, *Hydrodictyon*, arose from granules that were present on (*an*) the walls of the old tubes (1806, p. 3). He did not give particulars of the original positions of the granules, but what he saw were probably the pyrenoids, which are very evident in this plant. He derived the new cells of filamentous algae from the chloroplasts originally situated within pre-existing cells (1811, p. 6).

Kieser's writings (1814, pp. 105, 219) on cell-multiplication are not very explicit. He derived new cells from the small globules that are found in the *sève* contained in the intercellular spaces. He appears to have supposed that these globules originated within cells. Turpin (1829, p. 181) considered that Kieser's globules must have originated within cells; he allowed that they might perhaps develop into new cells in the intercellular spaces, in certain cases. As we shall see, however (p. 415), Turpin thought that new cells usually developed within pre-existing cells, and that no migration took place.

Endogeny without migration

According to this theory, which was supported by a formidable literature, small granules originate within a pre-existing cell (fig. 5); these granules enlarge at the expense of the contents of the pre-existing cell, until they touch one another and take on the usual characters of cells. The cell-wall of the mother-cell eventually disappears.

Grew's remark, quoted above (p. 409), that 'the greater Pores have some of them lesser ones within them', suggests that he may have envisaged this as a possible method of cell-multiplication. The theory, however, was first put forward in concrete form by Sprengel, in his account of the germination of the bean-seed (1802). He gives an illustration of cells of the cotyledon, with small granules or vesicles within them (his plate I, fig. 2), and he remarks, 'The small vesicles that still float in the fluid of the cell seem to have the

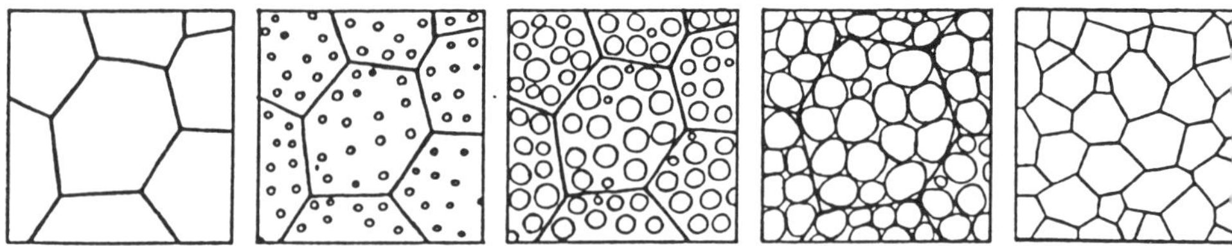

FIG. 5. Diagram of endogeny without migration.

character of future cells, and perhaps will become transformed into them subsequently' (p. 90). The granules or vesicles were actually starch-grains, the nature of which was not in the least understood at the time. In his book published in 1811, L. C. Treviranus also derived new cells from the granules contained inside the cells of the cotyledons of beans and peas (p. 4); in this later work he does not dogmatize as to the place of origin of these granules. Kieser (1814, p. 219) seems to have thought that the small globules that originate within cells and are the primordia of new cells, sometimes undergo their transformation without first passing out into the intercellular *sève*.

The theory of endogeny without migration flourished in the eighteen-twenties through the labours of Raspail and Turpin. From his studies of the germination of cereals, Raspail concluded that new cells arose from starch-grains, which enlarged within the cell that produced them until they touched one another, the mother-cell eventually bursting (1825, pp. 412–13). It is strange that the very man who discovered, by the iodine test, that these granules contain starch, should have been so misled about their fate. He elaborated his views in further communications. He imagined that each starch-grain contained within itself one or more globules, and that there were other exceedingly minute ones inside these (1827, p. 212, and his plate 2, fig. 22); this kind of *emboîtement* at the cellular level provided for repeated acts of cell-multiplication. He derived a whole leaf from a single cell, inside which two new cells arose endogenously and enlarged so as to fill all the space except what would become the midrib; globules arose within these two cells and enlarged to fill all the space except what would become the veins; and so on, till the final cellular structure of the mature leaf was achieved (1827, pp. 254–5, and his plate 4, fig. 4). He applied the same idea to the stems of plants

(p. 269) and to the tissues of animals (p. 304). He repeated his opinions on the endogenous origin of new cells, in his book on biochemistry (1833, pp. 85–86).

Turpin had completed his first paper on cell-multiplication by endogeny when he received Raspail's original communication on the subject (Turpin, 1827*a*, pp. 47–48). One may summarize his views, which he put forward at great length, by saying that he derived new cells from chromatophores and from colourless bodies which he believed to be of the same nature. He regarded *Pleurococcus naegelii*, so abundantly found on damp walls, as a solitary chromatophore. This plant, which he called a *globuline*, was for him a typical example of the most primitive organisms. The globuline contains within it a large number of smaller globulines, destined to reproduce the little organism (p. 25). Most plants, however, consist of colourless cells, which *contain* globulines; the latter are commonly green, but the starch-grains of the potato, for instance, are examples of white globulines (p. 42). Each globuline has the latent capacity to swell within its parent cell to form a new cell, losing its colour (if any) in the process (p. 41).

Turpin now investigated particular plants in the light of his ideas on cell-multiplication (1827*b*, 1828*a*). He derived new branches of *Enteromorpha* (which he called *Ulva*) from globulines situated within the cells. It is impossible to be certain of the real nature of these particular globulines; possibly they were zoospores. He gives a figure showing three branches of the plant that have originated from globulines all situated in a single cell (1828*a*; his plate 11, fig. 3).

In an extraordinary and tantalizing paper (1828*b*), Turpin mentions the great number of cases, both in simple microscopical plants and in the reproductive parts of higher forms, in which cells occur in aggregations of 2, 4, 8, or 16. He cites the pollen mother-cells of *Cobæa* (see below, p. 432). One would have thought that the idea of repeated binary fission would have forced itself on his imagination; but no, he thinks there is some unexplained tendency towards the germination of globulines in these particular numbers.

Turpin finally summarized his views at considerable length (1829), without adding anything of importance.

The ideas of Raspail were carried over into the embryology of animals by de Quatrefages (1834), in his study of the development of the pulmonate gastropods of fresh water. He thought that the early blastomeres or *globules* contained small similar bodies which grew and distended them, and that the process was repeated until a mass of cells had been produced, which took the form of the little mollusc (p. 115). Dumortier (1837), a student of the embryology of the same group of animals, appears to have held somewhat similar views. Misunderstanding the cleavage stages, he regarded the early embryo of *Limnaea* as being merely lobed and later facetted on the surface. He thought that cells appeared for the first time about a week after these stages. The cells that then appeared in the interior of the embryo he called *cellules primitives*. Inside each of these there arose eight or more *cellules secondaires*, indicated at first by certain *striatures obscures*. These secondary cells appear to have enlarged at the expense of material contained within the parent cell, until they filled it. (If, however, the striations divided the whole of the primary cell into secondary cells, the method should strictly be described as cell-division by partitioning (see below, p. 419).) The secondary cells subsequently enlarged until the primary cell burst, and only remnants of it were left. Only certain parts of the animal were formed of cells: the head was not (Dumortier, 1837, pp. 137 and 143–50, and his plate 4, fig. 16*a*).

The contributions of Schleiden to this subject must be considered in some detail, because they had a strong influence on contemporary opinion. He remarks (1838, p. 161) that new cells must either be formed outside the existing mass of tissue, or in its interior; if the latter, either in the intercellular spaces, or in the cells themselves; there is no fourth possibility (*quartum non datur*). The development of the plant occurs solely by the formation of cells within pre-existing cells and their subsequent expansion (pp. 163–5). He studied the development of new cells especially in the endosperm and pollen-tube. It may be remarked that he could scarcely have chosen an object of study more likely to lead him astray than the endosperm; for the development of a syncytium, with subsequent division into cells, does in fact bear some resemblance to the supposed process of endogeny without migration. The pollen-tube was almost as likely to lead to misinterpretation.

Schleiden gives a general account of the origin of new cells in these two situations. He describes the embryo-sack, in which the cells of the endosperm are to arise by endogeny, as a *Zelle* (p. 144). The first sign of impending cell-formation in the cytoplasm or *Gummi* of this cell, or of the pollen-tube, is the appearance of small mucus-granules. The nucleoli, larger and more sharply defined than the more numerous mucus-granules, are the next objects to appear. Schleiden calls them *Kernchen* (p. 145) or *Kerne der Cytoblasten* (p. 174); the descriptions and figures leave no doubt as to the correct interpretation of these names. It is unfortunate that in their translations of Schleiden's paper into English, both Francis (see Schleiden, 1841, p. 287) find Smith (see Schleiden, 1847, p. 238), overlooked what the original author said about the role of the nucleolus, apparently because they misread *Kernchen* as *Körnchen*; as a result, Schleiden's views have not till now been adequately represented to English readers. According to Schleiden himself (1838, p. 145), the nucleolus is the body round which the nucleus is formed, by the deposition in its immediate vicinity of a granular coagulum. (It is not clear whether the mucus-granules participate in this coagulum.) Schleiden called the nucleus the *Cytoblastus* (p. 139) or *Cytoblast* because he thought that its function was to produce the cell. According to his account (pp. 145–6) it grows larger, and a little blister, the rudiment of the future cell, appears on its surface. The contents of the blister are transparent. The appearance is rather like that of a watch-glass on a watch. The blister enlarges so as eventually to enclose the nucleus, except on one side. Its wall becomes stiffened into a jelly. When this process is complete, the blister has become a cell, the nucleus remaining enclosed in one place in its wall. The cell grows and assumes a regular shape as a result of the pressure of the other new cells surrounding it. The nucleus generally disappears after the cell has assumed its final form.

In his first paper on the cell-theory, Schwann (1838*a*) maintained that Schleiden's statements about the way in which cells multiply applied also to animals. He claimed to have found small cells within larger ones in the notochordal tissue and cartilage of the larvae of the spade-footed 'toad', *Pelobates*. In his book published the next year, he allowed that in animals new cells sometimes develop inside pre-existing ones, but he thought an exogenous origin much more usual (1839, pp. 45, 200, 203–4; see above, p. 412).

In his study of the earliest stages in the development of the rabbit, Barry concluded that two or more 'vesicles' (cells) originate within each pre-existing one (1839, p. 363). This is surprising, because he compares the early embryo with that of the frog; and he already knew, from the studies of Prévost and Dumas (1824), that in the latter animal the number of blastomeres increases by binary fission. Barry did not state clearly how he thought that cells multiply, although he mentioned the subject in several papers (1841, *a*, *b*, and *c*), which are unsatisfactory in more than one respect. He seems to have thought that the nucleus divides or fragments, and that each part of it grows to become a new cell.

Reichert considered that the blastomeres of amphibian eggs were formed endogenously within the uncloven egg, and smaller blastomeres endogenously within these, and so on: cleavage was merely the separation of blastomeres that were already present (1840, p. 7; 1841, p. 540).

Henle described what he thought to be a new cell arising endogenously round a nucleolus, within an existing (? human) cartilage-cell (1841, pp. 153–4 and his plate V, fig. 6). It is just possible that he was in fact observing a stage in cell-division.

Vogt considered that new cells sometimes arose within pre-existing cells, even on occasions within their nuclei (1842, pp. 126–7); but he thought that in animals exogeny was the more usual process (see above, p. 412). He regarded the nucleolus of the egg as a cell embedded in another cell, the nucleus, itself embedded in a third, the yolk (1842, p. 18).

A curious misapprehension prevented Kölliker from being among the first to understand the true nature of blastomeres. In his researches on the development of nematodes (1843), he got the fixed idea that the cells of the later embryo originate from what were really the nuclei of the earlier stages. These nuclei he called *Embryonalzellen*, to emphasize his opinion of their nature, and their nucleoli he regarded as nuclei (*Kerne*) (pp. 101–2). In some cases, in his view, there was no cleavage, but only a multiplication of the *Embryonalzellen*. Each of these produced two new small ones endogenously within it, and dissolved to set them free; the same process then happened repeatedly until all the numerous cells of the later embryo had been produced (p. 79). In other cases (e.g. in what he called *Ascaris nigrovenosa* (presumably another name for what is now called *Rhabdias bufonis*)), he described cleavage clearly enough, and gave excellent figures of it (his plate VI, figs. 21–23); but he totally misunderstood it. He knew that the blastomeres multiplied by division, but he was evidently not much interested in them, for he regarded them as mere spherical conglomerates of yolk-granules (pp. 105–6) round the all-important *Embryonalzellen*, which were going to multiply and produce the definitive cells.

In his general account of the multiplication of cells, published the next year in his book on the development of cephalopods (1844, pp. 141–57), he called the cells of adult animals *secondäre Zellen*, while nuclei were for him *primäre Zellen*. The nucleolus was the *Kern* of the primary cell. A blastomere was not a cell but an *Umhüllungskugel*. His views may be translated into modern terms as follows. In the early embryo, the nucleus of each blastomere ordinarily gives rise to two nuclei, by aggregation of material round the nucleoli. A substance which may be either granular or homogenous aggregates round each nucleus; the two aggregates separate from one another by a process involving the division of the original blastomere into two. This process continues until at last definitive cells are produced by the formation of cell-walls; these appear either round blastomeres, or else round nuclei. Kölliker also thought that a definitive cell might produce daughter-cells endogenously, apparently in the cytoplasm, and then degenerate to set them free. He also thought that new cells might arise within a mass of material formed by the fusion of cells.

Kölliker denied specifically that cells multiply by division. Having at last (1845) adopted a more acceptable nomenclature, he remarked without equivocation, 'Nothing whatever is known of a division of animal cells. Nuclei and cells multiply by endogenous procreation, nucleoli (*Kernchen*) by division' (p. 82). These are strange words from one who had observed cleavage so accurately.

In his old age, Kölliker claimed that in his book on the development of cephalopods (1844), he had made it very probable that all cells are the direct descendants of the blastomeres (Koelliker, 1899, p. 198). The truth is that in the book to which he refers he gave a confused and erroneous account of the way in which cells multiply, while the actual facts, as we shall see (pp. 430–1), had already been revealed by Bergmann in 1841–2.

J. Goodsir (1845, p. 2) regarded the nucleus as the source of successive broods of new cells, which grew within the mother-cell. It is not clear, however, whether he thought the new cells remained within the mother-cell or escaped from it. Goodsir attributed the discovery of the method by which cells multiply to Barry.

Beale (1865, pp. 241–2) appears to have been the last exponent of endogeny, though his remarks on the subject of the multiplication of cells are difficult to understand. He derived new 'elementary parts' (as he called cells) from minute particles, present (it would seem) within pre-existing cells. These particles enlarged, and meanwhile other similar particles might arise within them and also grow, and so on. This would be a clear example of endogeny without migration, but apparently the whole mass might divide and subdivide. When Beale reached these rather elusive conclusions, however, others had already discovered how cells actually multiply.

The Discovery of Cell-division

Preliminary remarks

About the middle of the nineteenth century there occurred a profound change in the beliefs of biologists about the way in which cells multiply. This change cannot be more dramatically recorded than in two extracts from the writings of Virchow. The first is from an original paper published in 1849. The second is the corresponding passage in a book of his collected papers, published seven years later. In the following translation an attempt is made to reproduce the style of Virchow's early writings, which are reminiscent of Oken's *Naturphilosophie.*

'The cell, as the simplest form of life-manifestation that nevertheless fully represents the idea of life, is the *organic unity*, the indivisible living One' (1849, p. 8).

'The cell, as the simplest form of life-manifestation that nevertheless fully represents the idea of life, is the *organic unity*, the divisible living One' (1856, p. 22).

In a note added to the second publication (p. 27), Virchow tries to persuade us that when he wrote *untheilbare* in 1849, he used the term in a philosophical rather than a scientific sense. It is difficult to accept this. There had, in fact, been a revolution in thought. It is the purpose of the rest of this paper to tell the story of this revolution and of the events that led up to it.

The early cytologists drew sharp distinctions between various methods of cell-division that seem to us to be very similar in all essential points. So sharp did these distinctions appear to them, however, that they would even describe cell-division while denying that cells ever divided. The difficulty is really verbal. They concentrated their attention on the cell-*wall*: this was for them the cell. If the wall did not divide, the cell did not divide, whatever might happen to its 'contents'.

Four methods of cell-division are illustrated diagrammatically in fig. 6.

In *cell-division by partitioning*, a thin membrane appears across the middle of a cell. It thickens and is seen to be a double partition, continuous with the pre-existing cell-walls. A single cell has become two cells, each of half the original volume. These grow.

In *cell-division with constriction of the cell-wall*, the latter bends inwards on all sides near the middle of the cell; a continuation of this process results in the division of the whole cell, including its wall. The two new cells grow. This was regarded as genuine cell-division by the early cytologists. It was called *Theilung durch Abschnürung.*

In *cell-division with formation of entirely new cell-walls*, the protoplasm divides into two or more parts inside the wall of the pre-existing cell. Each of these parts grows and acquires a complete new wall of its own, while the original wall disintegrates. This was the *Zellenbildung um Inhaltsportionen* of the early German cytologists. This name was given because the 'cells' (actually the

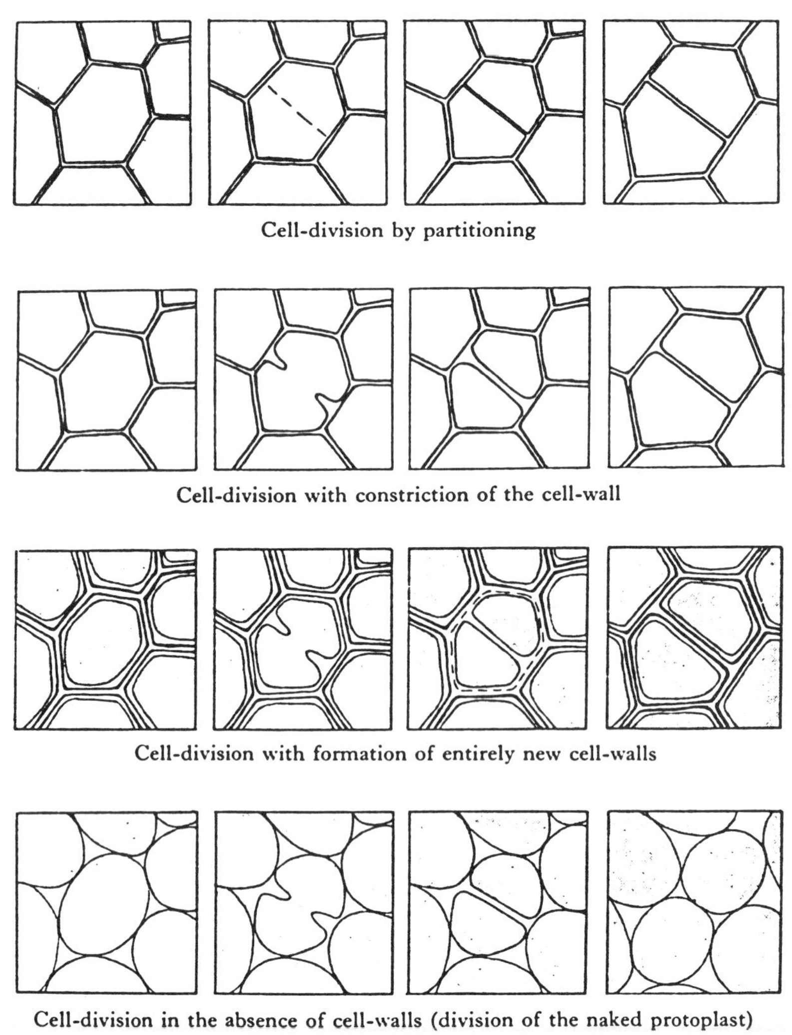

Cell-division by partitioning

Cell-division with constriction of the cell-wall

Cell-division with formation of entirely new cell-walls

Cell-division in the absence of cell-walls (division of the naked protoplast)

FIG. 6. Diagrams of methods of cell-division. In the two lower diagrams the protoplasts are shaded.

cell-walls) were formed afresh round portions of the cell 'contents' (protoplasm). In some cases the protoplasm divided up into numerous bodies that did not touch one another nor the wall of the mother-cell, and a new cell-wall was formed *separately* round each of these bodies. Since these new cell-walls were 'free' from each other, the name *freie Zellbildung* was used.

Cell-division with the formation of entirely new cell-walls shows a certain degree of resemblance to endogeny without migration. In endogeny, however, the new cells were supposed to originate as minute bodies that grew within the cytoplasm of the mother-cell, while in fact, of course, cells are first formed by the process of division, and growth is subsequent to this.

The endosperm is the classical site for the study of cell-division of this third type.

In *cell-division in the absence of cell-walls* (*division of the naked protoplast*) a furrow appears round the middle of a protoplast that has no cell-wall, and deepens until division is complete; the two resulting protoplasts then grow. This method of cell-division could not be envisaged until it was discovered that the cell-wall was not a necessary attribute of the cell. The history of that discovery was related in Part III of this series of papers (Baker, 1952).

Various different lines of research led up to the discovery that cells multiply by division into two or more parts. The chief of these were studies of protists, filamentous algae, and cleaving eggs. It would be possible to relate the story by concentrating first on examples of cell-division by partitioning, then on examples of division with constriction of the cell-wall, and so on; but the differences between the four methods result from such an unwarrantable overstressing of the cell-wall, that an unsatisfactory history would result. A far more logical arrangement will be to take each line of research separately. We shall begin with the results of researches on protists, for it was among these organisms that the process of cell-division was first witnessed.

The multiplication of protists

Leeuwenhoek saw ciliates coupled in pairs on several occasions (1681, p. 57; 1694, p. 198; 1697, p. 36; 1704, p. 1311). He interpreted the process in every case as one of copulation. Once he saw them actually come together in pairs *in conspectu meo* (1697, p. 36), and he must therefore have witnessed a stage in conjugation; but he does not give sufficiently accurate descriptions to make it certain whether he witnessed a stage in division on one or more of the other occasions. It is possible that he did (see especially 1694, p. 198). He himself, however, had no idea that ciliates multiply by division. He thought, on the contrary, that they reproduced by minute round particles (1697, p. 36), which in fact were presumably food-vacuoles (see also 1681, p. 56).

FIG. 7. The earliest illustration of a protist in division. (Anon., 1704, plate opposite p. 1329, fig. G (c).)

The first figure of a ciliate in division was given by an anonymous contributor to the Philosophical Transactions of the Royal Society (Anon., 1704; see fig. 7 in the present paper). The author himself did not regard this as a stage in multiplication by division. On the contrary, he compared the appearance with that of flies in copulation (pp. 1368–9). From what he says, it is quite possible that he saw ciliates in stages of both conjugation and division.

Joblot seems also to have seen a stage in the division of an unidentifiable ciliate (1718, plate 2, fig. 5). Like the anonymous writer, he regarded it as representing two individuals 'accouplées' (1718, part 2, p. 14), and indeed he appears to have seen stages in conjugation (his plate 2, fig. 1, and plate 3, fig. 9).

The first person to witness the process of multiplication of a protist by division was Trembley. He saw it in 1744 in the colonial vorticellid *Epistylis anastatica* and in *Stentor* (Trembley, 1746), and later in *Carchesium* and *Zoothamnium* (1748). I have described these discoveries in detail elsewhere (Baker, 1952*b*, pp. 103–12). It must suffice to say that Trembley's studies of these organisms, carried out with extraordinary care and accuracy, established for the first time the method of multiplication of Protozoa and provided a

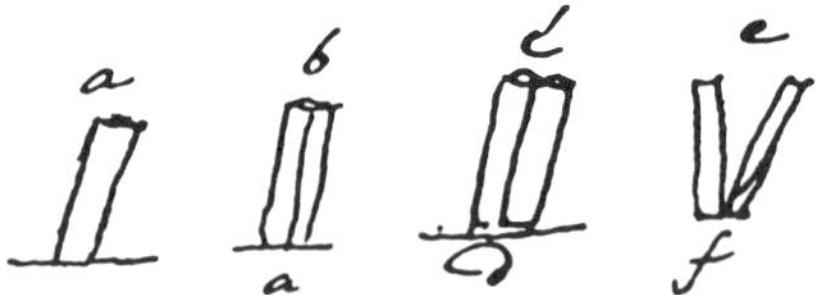

FIG. 8. The earliest illustration of cell-division. Trembley's sketch of the diatom *Synedra* dividing into two. (Trembley, 1766, folio 330.)

firm basis for disbelief in their spontaneous generation. The ciliates, however, can scarcely be considered as cells, in the sense in which that word is being used in this series of papers, on account of the highly polyploid nature of the macronucleus (see Baker, 1948*b*); and although this early work of Trembley's paved the way for the understanding of cell-division, yet it was not an investigation of cell-division itself. We shall therefore not pursue the subject of the reproduction of ciliates here, beyond remarking that Spallanzani, who corresponded with Trembley, also saw stages in their multiplication by division (Spallanzani, 1776, part 1, pp. 160 and 174–5, and his plate I, fig. VII, and plate II, figs. XIII and XIV).

Meanwhile Trembley himself had seen actual cell-division in the sessile, rod-shaped, fresh-water diatom, *Synedra*. I have described this discovery in detail elsewhere (Baker, 1951; 1952*b*, pp. 155–8). Trembley's sketch of the process is here reproduced in fig. 8. He noticed that a line appeared along the length of the organism, and became more conspicuous; then the whole object appeared to become a little wider, and the line was seen to be a groove; the parts on each side of the groove rounded themselves off from one another, and the previously single body was then seen to be double; finally the two halves of the originally single body diverged from one another at the unattached end. Trembley described this process first in a letter to a friend (1766, folio 330), and then, much later, in a book intended for the education of children (1775, vol. 1, pp. 293–7); in the interval another friend, Bonnet, had named the organism the *Tubiforme* and published the main facts of Trembley's discovery in his *Palingénésie philosophique* in 1769 (vol. 2, pp. 99–102). Trembley also observed cell-division in a stalked diatom, named by Bonnet the *Navette*;

this was almost certainly *Cymbella* (see Bonnet, 1769, vol. 2, pp. 104–5; Trembley, 1775, vol. 1, p. 297). Neither Trembley nor anyone else in his time realized that such organisms as *Synedra* and the component individuals of a *Cymbella* colony were cells.

Certain of Gleichen's figures suggest that he may have seen stages in the multiplication of the non-ciliate Protozoa of infusions, but the drawings are not clear enough to establish this (1778; see, e.g., his plate XVII, figs. C III and D III).

O. F. Müller described a dividing specimen of the desmid *Closterium*, which he called *Vibrio lunula* (1786, p. 57). His illustration is reproduced here in fig. 9. He also showed stages in the longitudinal division of a little organism found in stale sea-water, which may possibly have been a flagellate (his plate VIII, figs. 4–6).

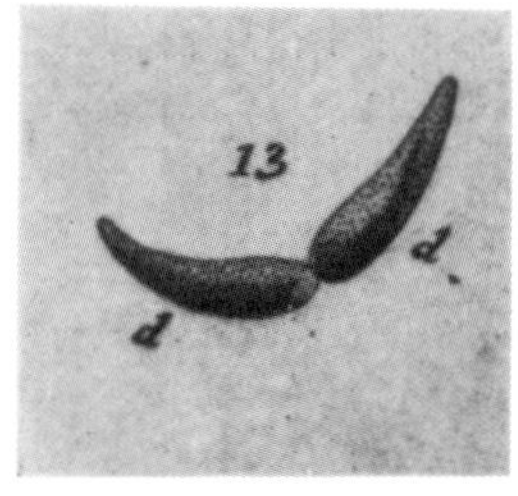

FIG. 9. O. F. Müller's figure of *Closterium* dividing into two. (Müller, 1786, plate VII, fig. 13).

The first person to describe the division of a protist with full realization that the process was one of cell-division, was Morren (1830). He describes *Crucigenia quadrata* as being ordinarily composed of *cellules* united in fours (see fig. 10). He describes the division of a single cell to form four, and of the four to form 16 (pp. 415–22). Later Morren saw stages in the multiplication of *Closterium*. He describes the extension inwards of a circular plate that makes a partition across the organism, which becomes jointed in this region; *déhiscence* then takes place (1836, p. 274).

Meanwhile Ehrenberg had started his celebrated researches. He described and figured an *Actinophrys* in the process of division (1830, p. 96); his illustration, here reproduced in fig. 11, A, suggests that the specimen was genuinely *Actinophrys*, not *Actinosphaerium*. Later (1832, p. 178) he saw a series of stages in the multiplication of *Euglena acus* by longitudinal fission (see fig. 11, B in the present paper). In his book on *Die Infusionsthierchen* (1838) he described and figured a number of examples of the multiplication of flagellates by division; for example, *Polytoma* (p. 25 and his plate I, fig. XXXII), *Pandorina* (p. 54 and plate II, fig. XXXIII), and *Glenodinium* (p. 257 and plate XXII, fig. XXII).

Nägeli described and figured the multiplication of certain diatoms by division (1844, plate I, figs. 1–6).

Cell-division in simple filamentous algae

The simplicity and immobility of most filamentous algae made them particularly suitable objects for the discovery of the way in which cells multiply.

In his careful researches on fresh-water algae, Vaucher at last succeeded in observing the germination of a zygote of *Spirogyra* (which he called *conferva jugalis*). He describes (1803, p. 47) how the cell-wall of the zygote (*grain*) opens at one end; a sack extends from it and begins to elongate into a tube.

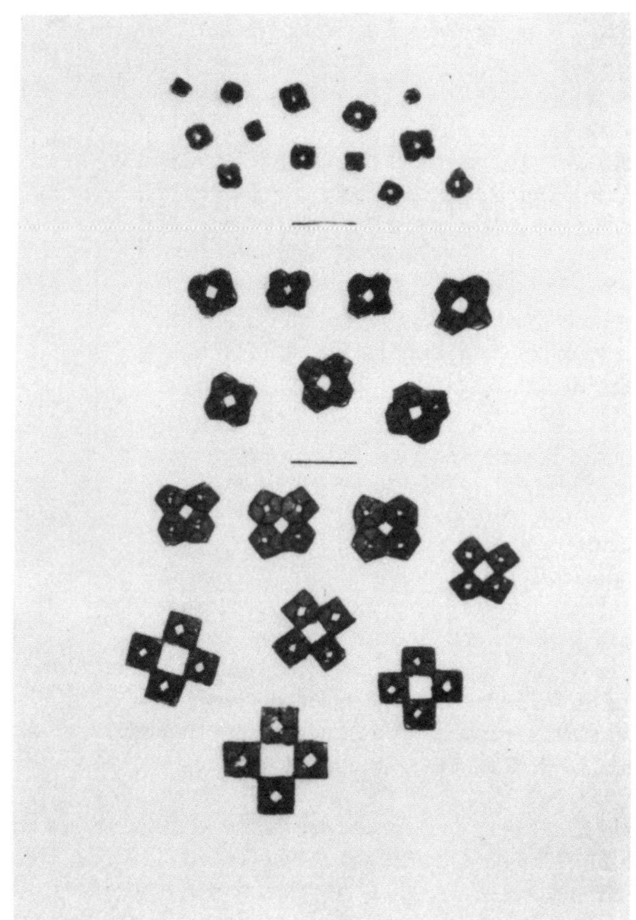

FIG. 10. Morren's figures of cell-division in *Crucigenia*. (Morren, 1830, plate 15, figs. 3–5.)

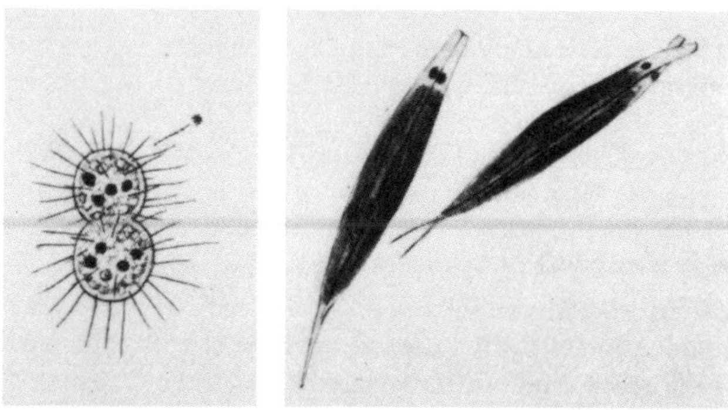

FIG. 11, A

FIG. 11, B

FIG. 11. Ehrenberg's figures of unicellular organisms in division. A, *Actinophrys sol*. (Ehrenberg, 1830, plate II, fig. IV (6).) B, *Euglena acus*. (Ehrenberg, 1832, plate I, fig. III (*b*, *c*). The plate (not the text) is dated 1831.)

He notes how the partitions (*cloisons*) between the cells (*loges*) appear: first one, then two, then many, until finally the tube resembles the plant that gave birth to it. His illustration is reproduced here (fig. 12). In his plate X, fig. 3, Vaucher shows stages in the germination of another alga; the number of partitions is seen to increase. The book contains nothing about the *binary* fission of the cells of any alga.

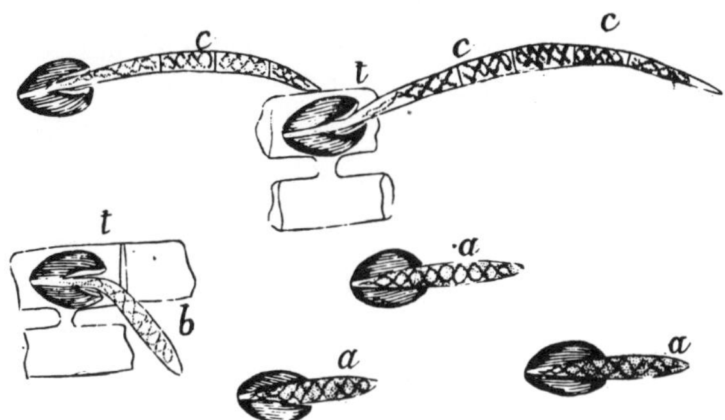

FIG. 12. Vaucher's figure showing new partitions between cells in a young *Spirogyra*. (Vaucher, 1803, plate IV, fig. 5.)

We have seen (p. 415) that Dumortier (1837) was mistaken about the way in which the cells of the embryo of *Limnaea* multiply. Five years previously, however, he had made an important contribution to the study of cell-multiplication in filamentous algae. He describes carefully (1832, pp. 226–7) how an extension inwards of the internal part of the cell-wall 'tends to divide the *cellule* into two parts'. He discusses whether the dividing wall or *cloison* is from the start double. He does not decide the question, but he says that in later stages it is certainly double in the conjugate filamentous algae. He supposed that cell-division was restricted to the cell at the extremity of a filament (see fig. 13 in the present paper). He remarks (p. 228) that new cells cannot originate from globules floating in the intercellular spaces, because some plants, such as the ones he was studying, have no such spaces.

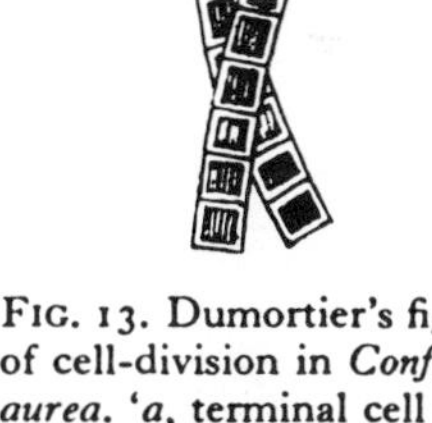

FIG. 13. Dumortier's figure of cell-division in *Conferva aurea*. '*a*, terminal cell that elongates more than the lower ones; *b*, the same divided into two parts by the formation of a median partition.' (Dumortier, 1832, plate X, fig. 15.)

Mohl's celebrated paper *On the multiplication of plant-cells by division* (1837) was first made public in the form of an inaugural lecture on his appointment as Professor of Botany at Tübingen in 1835. Like Dumortier, he studied filamentous algae (see fig. 14). His work on these organisms marks a turning-point in the history of the study of cell-multiplication, but he himself wrote with charming diffidence. 'Furthermore', he remarks, 'many appearances that I have observed in the various species of *Zygonema* [actually *Spirogyra*] make it seem to me more than likely that in

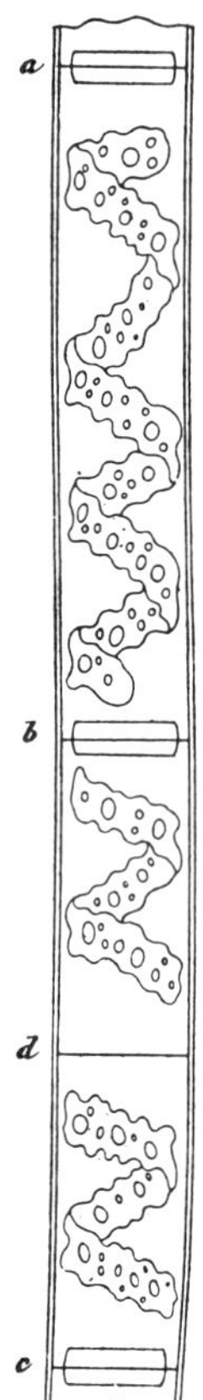

FIG. 14. Mohl's figure of cell-division in *Spirogyra*. The cell wall between the two newly formed cells is at *d*. (Mohl, 1837, plate I, fig. 8.)

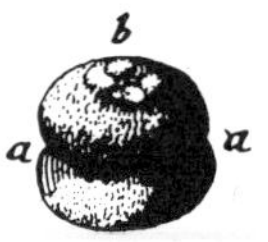

FIG. 15. Swammerdam's figure of a frog's egg, possibly at the 2-cell stage. (Swammerdam, 1737–8, vol. 2, plate XLVIII, fig. v.)

these plants also the individual cells possess the capacity to divide themselves in the middle by a partition-wall formed subsequently. . . . The observations cited above will suffice to prove that the increase of cells by division is not an altogether rare phenomenon among the Confervae' (pp. 29, 30).

Meyen (1838, p. 345) described the multiplication of the cells of certain filamentous algae by the process called in the present paper 'cell-division with constriction of the cell-wall'. He referred to this process, and also to cell-division by partitioning, by the name of *Theilung*.

Cleavage, and the recognition of blastomeres as cells

The study of the cleavage of the eggs of animals played an important part in convincing biologists that cells multiply by division. It was fairly easy to show that blastomeres did so; the difficulty was to discover that they were cells.

Swammerdam is thought by some to have seen a 2-cell stage in the cleavage of the frog's egg. One of his illustrations is reproduced here (fig. 15). He had placed the egg in a special fluid intended to dissolve the jelly, and this had so distorted it that one can neither affirm nor deny that he saw a stage in cleavage. He wrote, 'Next I observed the whole of the little frog divided, as it were, into two parts by a very obvious furrow or fold' (1737–8, vol. 2, p. 813). He never saw cleavage occurring in a living egg, nor did he see the 4-cell or later stages of the process. His observations on the embryology of frogs were perhaps made while he was studying these animals at Leiden in 1661–3. Dr. A. Schierbeck, however, who has made a careful study of the life of Swammerdam, thinks it probable that these observations were made in 1665. They were first published in the *Biblia naturae* long after his death.

It was stated by Bischoff (1842*a*, p. 46) that de Graaf (1672) saw a 2-blastomere stage in the rabbit. This is not true. De Graaf examined the Fallopian tubes and uteri of rabbits at various intervals after coition. He neither describes nor illustrates any stage in cleavage. From the third day onwards for several days he saw what must actually have been blastocysts (and perhaps late morulae), gradually increasing in size (pp. 313–14, and his plate XXVI, figs. 1–5). He saw no indication of the constituent cells.

Rösel von Rosenhof (1758) studied the development of several species of Anura. It is possible that he saw the 2-cell stage in the tree-frog, *Hyla arborea* (see his plate X, fig. 5, and p. 43). He also gives figures suggesting that he saw the first cleavage-furrow in *Rana temporaria* (plate II, figs. 9 and 10), but the accompanying text (p. 7) shows that the embryos were too old for this to be possible, and the drawings presumably represent neurulae.

Roffredi (1775) saw cleavage-stages in the free-living nematode *Rhabditis*. He figured the nuclei, but evidently did not understand what he was observing, for he did not notice the boundaries of the blastomeres (see Baker, 1949).

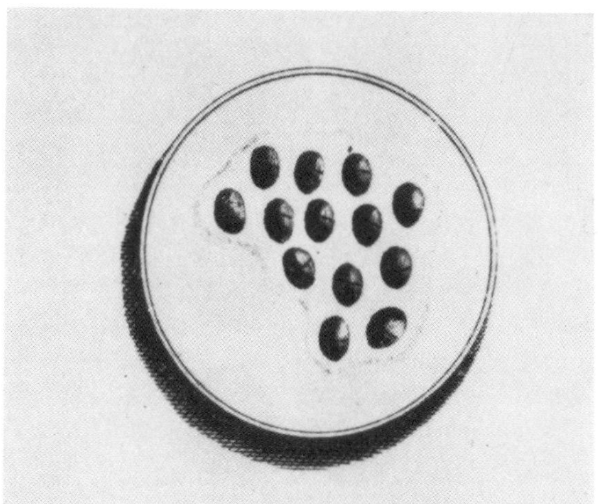

FIG. 16. Spallanzani's figure of the 4-cell stage in the development of the toad. (Spallanzani, 1780, plate II, fig. XI.)

It seems almost certain that Spallanzani (1780) saw the 4-cell stage in the toad (*Bufo*). His illustration is reproduced here in fig. 16. He calls the furrows *solchetti* (vol. 2, p. 25), and compares the appearance to that of the cupule of a chestnut when it has begun to split into its four lobes. He also gives what might be thought on casual reading to be a description of a 2-cell stage in the green frog (vol. 2, p. 13), but the *solco* was here actually the neural groove, not a cleavage-furrow. Spallanzani did not understand the process of cleavage. Indeed, he was so prejudiced by his belief in the actual pre-existence of later stages in the unfertilized egg, that he could not give any adequate account of the occurrences involved in development.

The process of cleavage, as it occurs in the living embryo, was first described by Prévost and Dumas in 1824. Their observations were made upon the eggs of what they called the *Grenouille commune*. This appears in fact to have been *Rana esculenta*, in which species the stages of cleavage are easier to observe than in *R. temporaria*. It is not clear whether they realized that the cleavage-furrows (*lignes* or *sillons*) went so deep as actually to divide the egg. They remark that the egg was soon *divisé* into two very pronounced *segmens* (p. 111), but from what they say farther on it is clear that at the stage to which they refer the furrow had not yet reached even the surface of the lower pole of the egg. In general, their observations were remarkably accurate. One of them observed and sketched what was occurring, while the other wrote down a short description. The later stages of cleavage occurred so rapidly that the authors could only compare them to the dissolving views (*changemens à vue*) seen at the theatre. Some of their drawings are here reproduced in fig. 17. They had no idea that the *segmens* were cells.

Prévost and Dumas noted the resemblance of the upper pole of the egg to

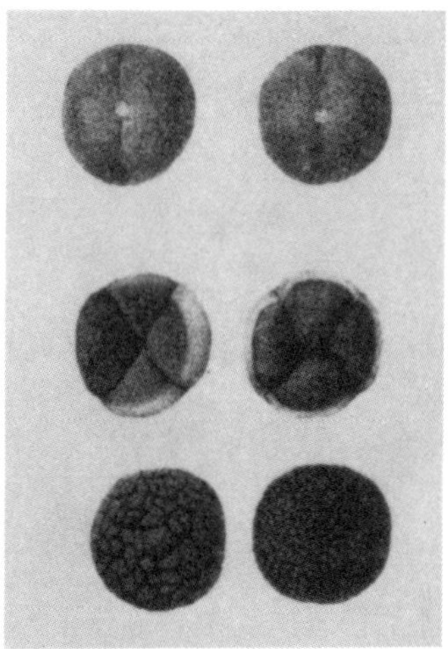

FIG. 17. Some of Prévost's and Dumas's figures of the cleavage of the frog's egg. (Prévost & Dumas, 1824, plate 6, figs. C, D, G′, G″, M, and N.)

a raspberry, at a late stage in cleavage (p. 112). As we shall see, later observers repeatedly noted the resemblance of such stages to raspberries, blackberries, or mulberries, in the development of various animals.

Cleavage stages in *Unio* were seen and figured by Carus (1832, pp. 44–45 and his plate 2, figs. I, III, X, XI), but he did not understand the origin or nature of the blastomeres.

It has already been mentioned (p. 415) that de Quatrefages (1834) saw the blastomeres or *globules* of various pulmonate gastropods of fresh water and thought that they multiplied by endogeny. He remarks (p. 109) that in the development of a species of *Limnaea*, the *globules* have formed a *tissu cellulaire* by the sixth day. The meaning of this is unfortunately not clear. We saw in Part I of this series of papers (1948, p. 112) that the expression *tissu cellulaire* was formerly used in a sense that seems very strange to us today, and de Quatrefages does not say distinctly that one globule represents one cell.

The cleavage of the eggs of frogs (*Rana temporaria* and *R. esculenta*) was re-investigated by von Baer (1834). The great embryologist thought that Prévost and Dumas had regarded the process as one of mere furrowing without actual division (von Baer, 1834, p. 481; 1835, p. 6). He made it perfectly clear that in fact the furrows actually divide the egg into discontinuous parts that are only pressed against one another (1834, p. 487). He used the word *Theilung* to describe the process. He compared the embryo at one stage to a blackberry and at a later stage to a raspberry (p. 493). He was thus the first to note the resemblance that was later perpetuated in the word *morula*. (The Romans used mōrum for both blackberry and mulberry. The diminutive form *morula* is a modern invention. The Latin word *mŏrula* meant a short delay.)

Rusconi gave excellent figures of cleavage in the *Wassersalamander* (1836*a*, his plate VIII, figs. 1–8) and of partial cleavage in the cyprinid fishes, *Tinca* and *Alburnus* (1836*b*, his plate XIII).

The cleavage of the egg by the formation of furrows had so far only been observed in vertebrates. Von Siebold now reported the same process in several genera of nematodes (1837). He called the blastomeres *Dottertheile* and remarked that when they had become small through repeated cleavage, the embryo resembled a blackberry. He saw the nuclei in the blastomeres from the 6-cell stage onwards; he did not use the word nuclei, but generally called them *helle Flecke* (p. 212). He called the nucleus of the egg the *Keimbläschen* or *Purkinjesche Bläschen*, and noticed the presence in it of a nucleolus or *Keimfleck* (p. 209). These observations constituted a considerable step towards the recognition of blastomeres as cells.

Schwann regarded *Dotterkugeln* as cells (1838*b*), but it is not clear from the context whether he here refers to blastomeres or yolk-globules. He recognized the protoplasts of the blastoderm of the hen's egg during the first day of incubation as nucleated cells, and gave a good figure of them, showing the nuclei and nucleoli (1839, pp. 63–66, and plate II, fig. 6).

It has been supposed that Cruikshank (1797) may have seen cleavage-stages in the rabbit, but neither his description nor his figures support this opinion. Jones may possibly have seen a morula-stage in the development of the same animal, but cells cannot be clearly seen in his illustration (1837, his plate XVI, fig. 1), nor does he describe them. The first person to describe the blastomeres of a mammal was the Scottish physician, Barry, who had worked with Schwann in Berlin (see Barry, 1838, p. 302). Barry (1839) gave admirable illustrations of cleavage-stages in the rabbit; two of them are here shown in fig. 18. As we have seen (p. 417), he was mistaken about the way in which blastomeres multiply, but credit is due to him for the first clear recognition that blastomeres are cells, or 'vesicles', to use his own word; he equates his vesicles with the cells of Schleiden and Schwann (p. 360). He remarks of the blastomeres of the 4-cell stage, 'Some of these vesicles presented in their interior a minute pellucid space, which may possibly have been a nucleus' (p. 323). In a footnote he adds, 'Later observations strengthen this supposition, and enable me to extend it to vesicles in the succeeding stages. The nucleus was very distinct in each of the two vesicles occupying the centre of the ovum in fig. 105½' (see fig. 18 in the present paper). He remarks, 'The nature of the alterations which the germ undergoes immediately after the termination of the primitive changes now referred to, I do not know, not having carried my investigations beyond that period. It is probable that they consist chiefly in the formation of new vesicles' (p. 365). In a later paper (1840, p. 542) he refers to the blastomeres of the 2-cell stage as 'cells'. Barry mentions the resemblance of the embryo at a certain stage to a mulberry (1838, p. 324).

The fact that blastomeres are cells was recognized by Reichert in 1840, independently of Barry. From his study of the frog's egg, Reichert concluded that the blastomeres gave rise to the cells of the adult, and he followed this out for various tissues (1840, pp. 13, 19, 58; 1841, p. 540). As we have seen,

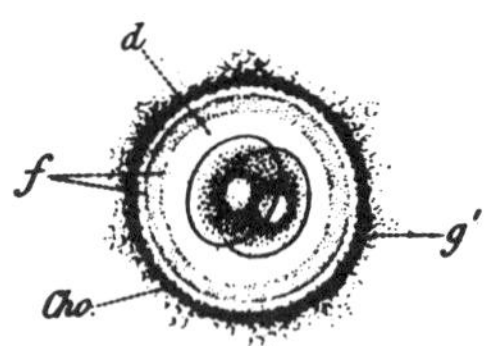

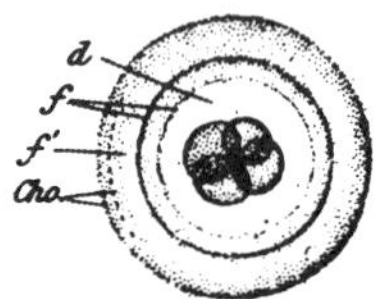

FIG. 18. Barry's figures of the 2- and 4-cell stages in the development of the rabbit. (Barry, 1839, plate VI, figs. 105½ and 106.)

however, he was entirely mistaken about the way in which bastomeres multiply (see above, p. 417).

The recognition by Barry and Reichert that blastomeres were cells constituted an important advance.

Bagge (1841) continued the work of Siebold by studying the early development of the eggs of nematodes. He followed cleavage in several species. In his figs. XI–XIX and XXI–XXIII (the latter wrongly labelled XXV–XXVII through the *Molestissima negligentia* of the engraver), he shows clearly, in the species he calls *Ascaris acuminata*, how the size of the cells is reduced by repeated division from the uncloven egg to the worm-shaped embryo. His great merit was his recognition (p. 10) that the large *vitelli partes* or blastomeres gave rise to the little *globuli* of the late embryo, by a process of repeated cleavage. It is unfortunate that when he used the word *cellulae*, he meant nuclei. He saw these in the blastomeres (e.g. in the 6-cell stage of *Strongylus*; see his fig. x). He noted the resemblance of the embryo at one stage to a blackberry or mulberry (one cannot say which, for he writes in Latin).

Bergmann of Göttingen, a student of the development of frogs and newts, was the first person who both understood the nature of cleavage and also recognized blastomeres as cells. He summed up his conclusions in words that show a restraint that is perhaps laudable. 'I may therefore state', he wrote, '*that the cleavage of the amphibian egg is an introduction to cell-formation in the yolk*. Indeed, I would even call it cell-formation, if the first, larger divisions of the yolk could unreservedly be called cells' (1841, p. 98). He compared his findings with those of Mohl in filamentous algae (see above, p. 425). He saw the *hellen Flecke* in the blastomeres, and thought that they might be nuclei.

Bergmann's slight hesitancy was caused by the absence of any resemblance between cleavage and the process of cell-formation as described by Schleiden. Although Bagge had not specifically homologized the *vitelli partes* and *globuli* with cells, yet Bergmann's hesitancy dissolved when he read the former's paper and found his own discoveries repeated in another group of animals. Bergmann now wrote of 'the identity of cleavage and cell-formation' (1842, p. 95). His two papers should form a landmark in the history of the cell-theory.

Vogt was one of the first to admit that Bergmann might be right, so far as frogs were concerned, but he denied that the latter's ideas were applicable to *Alytes* (1842, pp. 9, 25). Rathke (1842) recognized the blastomeres of *Limnaea* as cells: he saw within each its nucleus (*Kern*), and within the latter its nucleolus (*Kernkörper*). Bischoff (1842) gave excellent figures of cleavage in the rabbit (his plate III, figs. 21–26), but denied that the blastomeres of this animal were cells (p. 79). He was influenced mainly by the absence of cell-walls.

The truth now began gradually to be accepted. Reichert withdrew (rather half-heartedly) his idea that cleavage merely separated blastomeres that had already been formed endogenously, and allowed that it was the 'first act' of cell-formation (1846, pp. 274, 278). Kölliker was more explicit in his change of opinion. He made the important generalization that blastomeres always multiply by division, like infusoria, never by endogeny, as Reichert had supposed (Kölliker, 1847, pp. 12–13). Nearly a quarter of a century before, Prévost and Dumas had given a clear description of the cleavage of the frog's egg; but in the intervening years the theory of endogeny had taken root so firmly, that when blastomeres began to be regarded as cells, it was found hard to believe that they multiplied by division. Kölliker's generalization marked the end of the controversy on this subject. The credit, however, belongs to Barry, Reichert, Bagge, and especially Bergmann. Weldon (1898) was wrong in giving it to Kölliker.

The multiplication of other kinds of cells by division

Although the study of protists, filamentous algae, and blastomeres was of paramount importance for the discovery that cells multiply by division, yet quite a number of relevant observations were made from time to time on other objects. Grew may have had cell-division in mind when he remarked that some of the 'Pores' of the pith of plants were 'divided with cross Membranes' (1672; see above, p. 409); and Wolff, though he believed in exogeny as the usual method of cell-formation, yet allowed that partitions or *dissepimenta* were sometimes formed across the large cells of plants, with the production of smaller, included cells (1759, p. 21).

The division of the pollen mother-cells of *Cobaea scandens* (Polemoniaceae) into four young pollen-grains was observed by Brongniart (1827). His illustrations are here reproduced in fig. 19. The granules contained in the mother-cell, instead of forming a single mass, 'reunite in four perfectly distinct spherical masses, which float freely in the interior of the transparent utricle that contain them'. Each of these spherical masses 'continues to grow, and the membrane that covers it soon takes on a cellular aspect; the distended utricles that contain these globules in groups of four, split open' (p. 27). It is to be noticed that Brongniart did not describe binary cell-division.

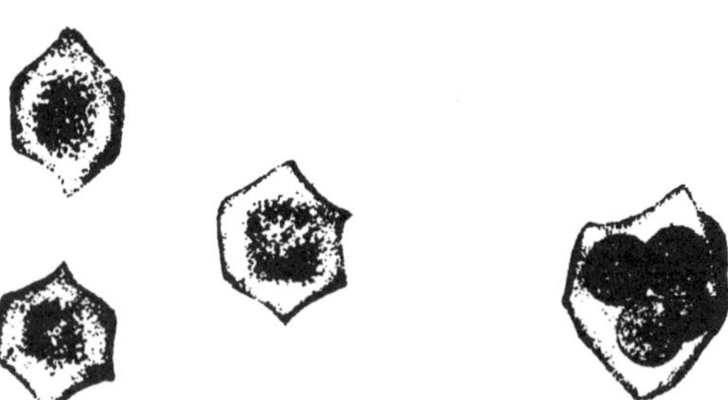

FIG. 19. Brongniart's figures of the division of the pollen mother-cells of *Cobaea scandens*. (Brongniart, 1827, plate 34, fig. 2 (E, F).)

Dumortier stated that all cells that are arranged in *rows* in the fronds of algae and in fungi, mosses, and Jungermanniales, multiply by partitioning in the same way as the cells of filamentous algae (1832, p. 229). Meyen, who was acquainted with Dumortier's findings, observed cell-division by partitioning in the developing lateral axes of *Chara*, and claimed to have observed it also in moulds (1838, pp. 339, 345). He would not use the name *Theilung* for those cases (e.g. endosperm-formation) in which new cell-walls are formed over the whole of the surface of the newly produced protoplasts; he referred to it instead as *Bildung der Zellen in Mutterzellen* (p. 346). Mirbel's *développement intra-utriculaire* (1835, p. 369) may perhaps have been cell-division. He seems to have recognized the varieties of the process that are called in the present paper 'partitioning' and 'cell-division with formation of entirely new cell-walls'.

Schwann himself allowed the possibility that in certain cases new cells might arise by partitioning of pre-existing cells (1839, pp. 5, 218), but he does not appear to have made any actual observations on this subject. Schleiden also equivocated slightly in a book (1842) published four years after his original communication. Though he still regarded endogeny as the standard method by which new cells arise in plants, and questioned the accuracy of Mohl's and Meyen's accounts of cell-division, yet he was clearly puzzled by observations he had made on the parenchyma of certain unspecified cacti. The cells were very regularly arranged, and he noticed here and there that one of them, though appearing single by its relation to the others, yet was clearly divided in two by a partition. This was very suggestive of multiplication by division; but Schleiden often noticed a *Cytoblast* on each side of the partition, and this allowed him to think it probable that even in this case, his own particular form of endogeny had been at work (pp. 267–9).

Remak (1841) would appear to have been the first person to observe a stage in cell-division in a many-celled animal, apart from cleavage. In the blood of a chick-embryo, in the third week of incubation, he observed pear-shaped cells, joined together in pairs by the stalks; each cell had a nucleus. He seems to have seen the remains of the spindle in the bridge between the two cells. He interpreted what he saw as a stage in cell-multiplication by division.

Vogt's observations on the notochord of the newt, *Triturus*, published the next year, were much more complete. He examined the cells at successive stages of larval development. 'The cell-wall bends inwards', he says, 'constricts, and thus at last divides into two halves, which are both exactly similar to one another and to the previously undivided cell, and they both continue their independent lives as cells' (1842, p. 128, and his plate 2, figs. 15 and 16; see also his pp. 46–47). It is strange that this clear description of the multiplication of cells by division should have been written by one who thought that cells ordinarily arise exogenously in a *Cytoblasteme* (see above, p. 412).

Valentin thought that blood-cells and other separate cells of multicellular animals in some cases multiplied by division (1842, p. 630). He instances the cells (*kernartigen Körperchen*) of the thymus gland of the embryo of the sheep; his figure (plate V, fig. 65) may indeed show actual stages of cell-division.

One of the most important contributors to our knowledge of cell-division was Nägeli (1844, 1846), but curiously enough, he himself would not allow that most of the processes he was studying constituted cell-division. He restricted the idea of division to *Abschnürung*, that is to say, to what is called in the present paper 'cell-division with constriction of the cell-wall'. In his earlier paper (1844, p. 97) he denied that this process was ever complete: a partition appeared before the constriction had become very deep. He writes of 'so-called' cell-division (p. 110). Later, however, he allowed the reality of complete *Abschnürung* in certain cases (1846, p. 60). By 'cell-formation' he meant the formation of cell-walls. For him, the cell-wall *was* the cell: if it did not constrict to nip the pre-existing cell in two, cell-division did not occur.

Although he concentrated so much attention on the wall of the cell, Nägeli by no means overlooked its *Inhalt* or protoplasm, and indeed he made some interesting observations on the multiplication of nuclei, to which it will be necessary to refer in the next paper in this series. He noticed that when cell-multiplication is about to occur, there is first of all an isolation or individualization of parts of the *Inhalt* (or, in modern terms, the protoplasm divides into two or more parts) (see fig. 20). A *Membran* or thin cell-wall then forms round each of the parts. In *wandständige Zellenbildung* this new cell-wall, from the moment of its appearance, is everywhere in contact with the wall of the original cell, except where the new *Inhaltspartien* are

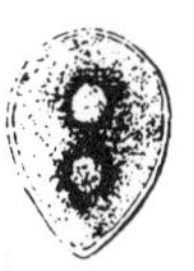

FIG. 20. Nägeli's figure of the division of a germinating spore of *Padina* (Phaeophyceae). (Nägeli, 1844, plate II, figs. 4 and 5.)

divided from one another; here a new partition is formed. This is a kind of cell-division by partitioning, in the terminology used in the present paper. In *freie Zellenbildung* the *Inhaltspartien* separate themselves entirely so that each is 'free'; a new wall is formed round each, and these walls are nowhere in contact with the original cell-wall (Nägeli, 1846, pp. 51, 60, 62; see p. 420 of the present paper).

The strangeness of Nägeli's papers is mainly verbal. When once one has grasped his use of words, it is clear that he made a massive contribution to our knowledge of the process by which cells multiply. He studied it in all the major groups of plants (other than bacteria). From his time onwards scarcely anyone could take seriously the contention that plant-cells multiply exogenously, and the foundation had been laid for a general understanding that the protoplast multiplies by division.

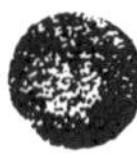  

FIG. 21. Reichert's figures of the division of the primary spermatocyte of *Strongylus auricularis* into 4 spermatids. (Reichert, 1847, plate VI, figs. 5–7.)

It will be remembered that Reichert was mistaken about the nature of cleavage (see above, p. 417). He made important observations, however, on the multiplication of the male germ-cells of the nematode *Strongylus* (1847). He identified the cells at the blind upper end of the tubular testis as 'elementary nucleated cells', and understood that a series of stages in spermatogenesis was displayed along the tube, the pear-shaped ripe spermatozoa being found at the opposite end. Working along the tube from the blind end, he saw stages in the division of the spermatogonia (pp. 101–2), and further along again he saw and figured what were evidently the meiotic divisions (pp. 110–14); these he compared with the processes of pollen-formation (which, as we have seen, had been observed by Brongniart). Some of his figures are reproduced here in fig. 21. Like Nägeli, Reichert could not escape from the ideas of the nature of the cell that were current in his time, and he described cell-division as *Zellenbildung um Inhaltsportionen.* The process whereby the *Inhalt* (protoplasm) divided into its *Portionen* (daughter-protoplasts) was evidently of secondary interest to him, for he was always looking for the formation of a *Zelle* (cell-wall) round a newly formed protoplast. Like Nägeli again, however, he by no means overlooked the *Inhalt*, and it will be necessary to revert to his work on *Strongylus* in the next paper in this series, in which the multiplication of nuclei will be considered.

Omnis cellula e cellula

'The origin of cells', wrote de Candolle in 1827, 'like everything connected with the origin of organisms, is a problem that it is absolutely impossible to resolve in the present state of knowledge' (p. 27). Twenty years later the

main facts had been discovered. Morren had recognized *Crucigenia* as a cell and followed its multiplication by division. Dumortier, Mohl, and Meyen had discovered how the cells of filamentous algae multiply. Dumortier and especially Nägeli (despite his misleading terminology) had established cell-division as the standard method of cell-multiplication in the main groups of plants. That the cleavage of eggs is cell-division had been revealed by the investigations of von Siebold, Barry, Reichert, Bagge, and Bergmann. Finally, cell-division had been demonstrated in notochordal and spermatogenetic tissues by Vogt and Reichert. It remained to realize that the process that had been studied and reported over and over again was the universal method by which cells multiply.

This realization, epitomized in the words *Omnis cellula e cellula* (surely one of the grandest inductions of biology), we owe almost entirely to two men, Remak and Virchow. The Latin phrase, however, does not exclude the origin of new cells by endogeny, and it must be remarked in passing that Raspail, Turpin, Schleiden, and Goodsir would perhaps have assented to it, if it had been put forward in their time. In fact, however, the phrase was introduced solely in reference to the origin of new cells by *division*.

Remak and Virchow had both been pupils of Johannes Müller, both were practical medical men as well as biologists, and both were in their thirties. In other respects they were very different. Remak was the typical research-worker. He carried out thorough investigations in the laboratory; he studied carefully the work of others, and made full acknowledgement of it; he wrote in a straightforward style and eschewed all fanciful ideas. Virchow, on the contrary, soared away in the manner of his predecessors in the school of *Naturphilosophie*, and left the reader guessing what the actual facts were that led him to his conclusion, and who discovered them.

Although both men published in 1852 and their papers cannot be exactly dated, yet the circumstantial evidence suggests that Remak was the first in the field. It will be remembered that he had observed a stage in the multiplication of blood-corpuscles by division in 1841 (see p. 433). In 1851, when writing on the cleavage of the frog's egg, he said that he must reserve to another paper his remarks on the transition from the cleavage-cells to the tissues, by repeated cell-division (p. 496). The promised article appeared in the following year.

By a careful review of the available evidence, but without adding new observations, Remak (1852) set out to explode Schwann's idea of the exogenous origin of cells and to set up in its stead a general theory of their multiplication by division. He remarks that the botanists no longer believe that cells arise outside pre-existing cells. To him, the extra-cellular origin of cells is as unlikely as the *Generatio aequivoca* of organisms (p. 49). Between the cleavage-cells there is no intercellular substance in which new cells could originate exogenously. Remak breaks loose at last from the domination of the vitelline membrane, which had so misled previous investigators. For them, the vitelline membrane *was* the cell; the protoplasm was its *Inhalt*. Thus, since the vitelline membrane did not change during cleavage, cell-division did not occur!

Remak plays down the *Dotterhaut*, remarking that it does not participate in the formation of the egg-cell. The protoplasm of the egg-cell passes over into that of the embryonic cells, and the nuclei of the latter are the derivatives of the nucleus of the first cell. Remak thinks it unlikely that new cells arise from extracellular substance even in diseased tissues. 'The statement that the cells of animals, like those of plants, have only an *intracellular* origin, seems to me to be a proposition established by a long series of reliable experiences' (p. 55).

Remak's elaborate study of the development of the chick and frog, recorded in his celebrated book (1855), convinced him of the correctness of the views he had expressed in 1852. Towards the end of the work, in a valuable general review of the cell-theory, he upheld division (*Theilung*) as the standard method of cell-formation. He is concerned once again by that bugbear of the cytologists of his time, the vitelline membrane, and points out that it is not always possible to distinguish with certainty between cell-membranes, thickenings of the outer parts of cells, and intercellular material (p. 174). He realizes that what is essential is the protoplasm, and this *divides*; 'all animal cells arise from the embryonic cells by progressive division' (p. 178). He is puzzled, however, by the fact that in certain rhabdocoels the germinal vesicles are formed in one organ, and the yolk in another; this makes him hesitate about saying unequivocally that the ordinary egg, with its yolk, is a cell. He understands clearly that the division of the egg by cleavage-furrows is not always complete.

It is impossible to tell whether Virchow had read Remak's paper of 1852 when he published his own in the same year. He makes no mention of Remak; but this is perhaps not very significant, for he mentions no one who had written on the multiplication of cells except Schleiden, Schwann, and Kölliker (and the latter only in connexion with the contractility of the vessels of the umbilical cord!). One does not know what were the chief facts that convinced him of the origin of cells from pre-existent cells by *Zertheilungen und Zerspaltungen* (1852, p. 377). Somewhat understating the case, he admits that his earlier definition of the cell as 'the indivisible living One' (see above, p. 419) was *nicht ganz richtig*; for *überall findet sich das Princip der Theilbarkeit, der Spaltbarkeit* (p. 378).

Like Remak, he returned to the theme in 1855. Like Remak, he pointed out that if cells did not arise from pre-existent cells, the state of affairs would resemble *Generatio aequivoca*. Unlike Remak, he uses strangely violent language in denouncing spontaneous generation as 'either pure heresy or the work of the devil'. He then proceeds to the great generalization. 'I formulate the doctrine of pathological generation, of neoplasia in the sense of cellular pathology, simply thus: *Omnis cellula a cellula*' (p. 23). It is noteworthy that in coining the aphorism, Virchow applies it to diseased tissues, and takes for granted that it applies to normal cells. It seems to have been Leydig who first put the phrase in its final form. Like Remak and Virchow, Leydig first of all denies the reality of *generatio aequivoca*. 'Observation knows only an *increase of cells from themselves*,' he proceeds, 'and the same validity might be ascribed to the proposition *omnis cellula e cellula* as to *omne vivum e vivo*.' This was the form of words adopted by Virchow in his statement on the subject in the *Cellularpathologie* (1859, p. 25). 'Just as we no longer allow', he wrote, 'that a roundworm originates from mucous slime, or that an infusorian or a fungus or an alga forms itself from the decomposing remains of an animal or plant, so also we do not admit in physiological or pathological histology that a new cell can build itself up from a non-cellular substance. Wherever a cell originates, in that place there must have been a cell before (*Omnis cellula e cellula*), just as an animal can only originate from an animal and a plant from a plant.'

If we deliberately overlook the strange style in which Virchow wrote his paper of 1852, we can see that he made a contribution towards the understanding that cells are derived from pre-existing cells by division. But if we compare his writings with Remak's, we cannot fail to recognize that the latter's must have had far more influence. Indeed, it does not seem likely that Virchow's writings by themselves would have had much effect upon opinion. Remak's paper of 1852 contains no catch-phrase, but it stands out as the first clear and solidly backed general statement of the way in which cells multiply.

It must be regretted that in later years Remak (1862) withdrew to some extent from the position he had adopted. No exception was definitely known to the rule that in normal tissues, cells multiplied by division: that he still allowed. But he now maintained that endogeny occurred in diseased tissues, and that in this process, pre-existing nuclei were not concerned. Further, he thought it probable that certain cells in normal tissues multiplied endogenously. He considered that spermatozoa originated within a mother-cell, and that merogony was a form of endogeny. The nuclei of certain cells, in his view, could not be traced back to the nuclei of the embryo. He instanced the star-shaped cells of connective tissue (presumably the fibroblasts), and the cells of the cutis and of the smaller branches of the blood-vessels of the frog.

These doubts must not be allowed to obscure the service that Remak had rendered to biology at the appropriate moment, ten years before. Nevertheless, we must guard against overestimating his contribution. Important though he was, Remak was not the discoverer of the way in which cells multiply. Trembley, Morren, Ehrenberg, Dumortier, Mohl, Meyen, Prévost, Dumas, von Siebold, Barry, Reichert, Bagge, Bergmann, Nägeli—these were the men whose discoveries had produced a situation in which a great generalization would be acceptable. Remak supplied it.

Acknowledgements

I am particularly grateful to Dr. Charles Singer for calling my attention to two important papers that I should otherwise have missed. I take the opportunity of adding Klein's *Histoire des origines de la théorie cellulaire* (1936) to the list of valuable works on the subject, given in the first part of this series of papers. My work on the cell-theory has continued to receive the support of Professor A. C. Hardy, F.R.S.

Correction. In part III of this series of papers (1952), I suggested that the word *coenocyte* appeared to be superfluous. Further consideration has led me to change this opinion. It seems desirable to have a special name for those syncytia that resemble single cells in shape (e.g. the binucleate components of mammalian liver). The word coenocyte will therefore be used in future parts of this series of papers when it is necessary to specify this particular category of syncytia.

REFERENCES

ANON., 1704. Phil. Trans., **23**, 1357.

BAER, K. E. v., 1834. Arch. Anat. Physiol. wiss. Med., (no vol. number,) 481.

—— 1835. Bull. sci. Acad. Imp. St. Pétersbourg, **1**, 4.

BAGGE, H., 1841. *Dissertatio inauguralis de evolutione Strongyli auricularis et Ascaridis acuminatae viviparorum.* Erlangae ex Officina Barfusiana.

BAKER, J. R., 1948*a*. Quart. J. micr. Sci., **89**, 103.

—— 1948*b*. Nature, **161**, 548.

—— 1949. Quart. J. micr. Sci., **90**, 331.

—— 1951. Isis, **42**, 285.

—— 1952*a*. Quart. J. micr. Sci., **93**, 157.

—— 1952*b*. *Abraham Trembley of Geneva, scientist and philosopher, 1710–1784.* London (Arnold).

BARRY, M., 1838. Phil. Trans., **128**, 301.

—— 1839. Ibid., **129**, 307.

—— 1840. Ibid., **130**, 529.

—— 1841*a*. Ibid., **131**, 193.

—— 1841*b*. Ibid., **131**, 195.

—— 1841*c*. Ibid., **131**, 217.

BEALE, L. S., 1865. Arch. of Med., **2**, 207.

BERGMANN, —, 1841. Arch. Anat. Physiol. wiss. Med., (no vol. number,) 89.

—— 1842. Ibid., 92.

BISCHOFF, T. L. W., 1842. *Entwicklungsgeschichte des Kaninchen-Eies.* Braunschweig (Vieweg).

BONNET, C., 1769. *La palingénésie philosophique, ou idées sur l'état passé et sur l'état futur des êtres vivans.* 2 vols. Genève (Philibert & Chirol).

BRONGNIART, A., 1827. Ann. des Sci. nat., **12**, 14.

BURDACH, K. F. (edited by), 1837. *Die Physiologie als Erfahrungswissenschaft.* 2nd edit. Vol. 2. Leipsig (Voss).

CANDOLLE, A.-P. DE, 1827. *Organographie végétale, ou description raisonnée des organes des plantes.* 2 vols. Paris (Deterville).

CARUS, C. G., 1832. *Neue Untersuchungen über die Entwickelungsgeschichte unserer Flussmuschel.* Leipsig (Fleischer).

CRUIKSHANK, W., 1797. Phil. Trans., **87**, 197.

DUMORTIER, B. C., 1832. Verh. kais. Leopold.-Carol. Akad. Naturf., **16**, 217.

—— 1837. Ann. des Sci. nat. Zool., **8**, 129.

EHRENBERG, [D.] C. G., 1830. *Organisation, Systematik und Geographisches Verhältnis der Infusionsthierchen.* Berlin (Akademie der Wissenschaften).

—— 1832. *Zur Erkenntnis der Organisation in der Richtung des kleinsten Raumes.* Berlin (Akademie der Wissenschaften).

—— 1838. *Die Infusionsthierchen als vollkommene Organismen.* Leipsig (Voss).

GLEICHEN, W. F., 1778. *Abhandlung über die Saamen- und Infusionsthierchen, und über die Erzeugung.* Nürnberg (Winterschmidt).

GOODSIR, J. (and J. D. S.), 1845. *Anatomical and pathological observations.* Edinburgh (MacPhail).

GRAAF, R. DE, 1672. *De mulierum organis generationi inservientibus tractatus novus.* Lugduni Batav. (ex Officina Hackiana).

GREW, N., 1672. *The anatomy of vegetables begun. With a general account of vegetation founded thereon.* London (Hickman).

—— 1682. *The anatomy of plants. With an idea of a philosophical history of plants. And several other lectures, read before the Royal Society.* Published by the author (place not stated).

HENLE, J., 1841. *Allgemeine Anatomie. Lehre von den Mischungs- und Formbestandtheilen des menschlichen Körpers.* Leipsig (Voss).

JOBLOT, L., 1718. *Descriptions et usages de plusieurs nouveaux microscopes, tant simples que composez* . . . Paris (Collombat).

JONES, T. W., 1837. Phil. Trans., **127**, 339.

KIESER, D. G., 1814. *Mémoire sur l'organisation des plantes.* Harlem (Beets).

KLEIN, M., 1936. *Histoire des origines de la théorie cellulaire.* Paris (Hermann).

KÖLLIKER [KOELLIKER], A., 1843. Arch. Anat. Physiol. wiss. Med., (no vol. number,) 68.

—— 1844. *Entwickelungsgeschichte der Cephalopoden.* Zürich (Meyer & Zeller).

—— 1845. Zeit. wiss. Bot., **1** (2), 46.

—— 1847. Arch. f. Naturges., **13**, 9.

—— 1899. *Erinnerungen aus meinem Leben.* Leipsig (Engelmann).

LEUWENHOEK [LEEUWENHOEK, LEUWENHOCK], A. [A. VAN], 1681. Phil. Collections (R. Hooke), **2**, 51.

—— 1694. Phil. Trans., **18**, 194.

—— 1697. *Continuatio arcanorum naturae detectorum.* Delphis Batavorum (Kroonevelt).

—— 1704. Phil. Trans., **23**, 1304.

LINK, D. H. F., 1807. *Grundlehren der Anatomie und Physiologie der Pflanzen.* Göttingen (Danckwerts).

MEYEN, F. J. E., 1838. *Neues System der Pflanzen-Physiologie.* 3 vols. Berlin (Spenersche Buchhandlung).

MIRBEL, —, 1835. Mém. Acad. Roy. Sci. Inst. France, **13**, 337.

MOHL, H., 1837. Allg. bot. Zeit., **1**, 17.

MORREN, C. F.-A., 1830. Ann. des Sci. nat., **20**, 404.

—— 1836. Ann. des Sci. nat. Bot., **5**, 257.

MÜLLER, O. F., 1786. *Animalcula infusoria fluviatilia et marina.* Hauniæ (Mölleri).

NÄGELI, C., 1844. Zeit. wiss. Bot., **1** (1), 34.

—— 1846. Ibid., **1** (3), 22.

PRÉVOST, —, and DUMAS, —, 1824. Ann. des Sci. nat., **2**, 100.

QUATREFAGES, A. DE, 1834. Ibid., **1**, 107.

RASPAIL, [F. V.], 1825. Ibid., **6**, 384.

—— 1827. Mém. Soc. d'Hist. nat. (Paris), **3**, 209.

—— 1833. *Nouveau système de chimie organique, fondé sur des méthodes nouvelles d'observation.* Paris (Baillière).

RATHKE, H., 1842. Neue Not. Geb. Nat. Heilk. (Froriep), **24**, 160.

REICHERT, K. B. [C. B.], 1840. *Das Entwickelungsleben im Wirbelthier-Reich.* Berlin (Hirschwald).

—— 1841. Arch. Anat. Physiol. wiss. Med., (no vol. number,) 523.

—— 1846. Ibid., (no vol. number,) 196.

—— 1847. Ibid., (no vol. number,) 88.

REMAK, [R.], 1841. Med. Zeit., **10**, 127.

—— 1851. Arch. Anat. Physiol. wiss. Med., (no vol. number,) 495.

—— 1852. Arch. path. Anat. Physiol. klin. Med. (Virchow), **4**, 375.

—— 1855. *Untersuchungen über die Entwickelung der Wirbelthiere.* Berlin (Reimer).

—— 1862. Arch. Anat. Physiol. wiss. Med., (no vol. number,) 230.

ROFFREDI, M. D. [*sic*], 1775. Journal de Physique (Obs. et Mém. sur la Physique), **5**, 197.

RÖSEL VON ROSENHOF, A. J., 1758. *Die natürliche Historie der Frösche hiesigen Landes.* Nürnberg (Fleischmann).

RUDOLPHI, K. A., 1807. *Anatomie der Pflanzen.* Berlin (Myliussischen Buchhandlung).

RUSCONI, M., 1836*a*. Arch. Anat. Physiol. wiss. Med., (no vol. number,) 205.

—— 1836*b*. *Ibid.*, (no vol. number,) 278.

SCHIERBEEK, A., 1953. Personal communication.

SCHLEIDEN, M. J., 1838. Arch. Anat. Physiol. wiss. Med., (no vol. number,) 137.

—— 1841. Sci. Memoirs, **2**, 281. (Translated from Schleiden (1838) by W. Francis.)

—— 1842. *Grundzüge der wissenschaftlichen Botanik nebst einer methodologischen Einleitung als Anleitung zum Studium der Pflanze.* Leipsig (Engelmann).

—— 1847. *Contributions to phytogenesis.* London (Sydenham Society). (Translated from Schleiden (1838) by H. Smith.)

SCHWANN, T., 1838*a*. Neue Not. Geb. Nat. Heilk. (Froriep), **5**, column 33.
—— 1838*b*. Ibid., **6**, column 21.
—— 1839. *Mikroskopische Untersuchungen über die Uebèreinstimmung in der Struktur und dem Wachstum der Thiere und Pflanzen.* Berlin (Sander'schen Buchhandlung).
SIEBOLD, K. T. VON., 1837. Article 'Zur Entwickelungsgeschichte der Helminthen' in Burdach, 1837, p. 183.
SPALLANZANI, —, 1776. *Opuscoli di fisica animale, e vegetabile.* 2 parts. Modena (Società Tipografica).
—— 1780. *Dissertazioni di fisica animale, e vegetabile.* 2 vols. Modena (Società Tipografica).
SPRENGEL, K., 1802. *Anleitung zur Kenntnis der Gewächse.* Halle (Kümmel).
SWAMMERDAM, J., 1737–8. *Biblia naturae; sive historia insectorum, in classes certas redacta.* 2 vols. Leydae (Severinum, Vander, Vander).
TREMBLEY, A., 1746. Phil. Trans., **43**, 169.
—— 1748. Ibid., **44**, 627.
—— 1766. Manuscript letter to Count Bentinck. Folio 330 of 'Correspondence of Count Bentinck, with his son Antoine, and his tutors, 1740–1765'. British Museum (ref. Egerton 1726).
—— 1775. *Instructions d'un père à ses enfans, sur la nature et sur la religion.* 2 vols. Genève (Chapuis).
TREVIRANUS, L. C., 1806. *Vom inwendigen Bau der Gewächse und von der Saftbewegung in demselben.* Göttingen (Dieterich).
—— 1811. *Beyträge zur Pflanzenphysiologie.* Göttingen (Dieterich).
TURPIN, P.-J.-F., 1827*a*. Mém. Mus. d'Hist. Nat. (Paris), **14**, 15.
—— 1827*b*. Ibid., **15**, 343.
—— 1828*a*. Ibid., **16**, 157.
—— 1828*b*. Ibid., **16**, 295.
—— 1829. Ibid., **18**, 161.
VALENTIN, G., 1835. *Handbuch der Entwickelungsgeschichte des Menschen mit vergleichender Rücksicht der Entwickelung der Säugethiere und Vögel.* Berlin (Rücker).
—— 1839. 'Uebersicht über Histogenese', contributed to Wagner, 1839, p. 132.
—— 1842. Article on 'Gewebe des menschlichen und thierischen Körpers' in Wagner, 1842.
VAUCHER, J.-P., 1803. *Histoire des conferves d'eau douce.* Genève (Paschoud).
VIRCHOW, R., 1849. *Die Einheitsbestrebungen in der wissenschaftlichen Medicin.* Berlin (Reimer).
—— 1852. Arch. path. Anat. Physiol. klin. Med. (Virchow), **4**, 375.
—— 1855. Ibid., **8**, 3.
—— 1856. *Gesammelte Abhandlungen zur wissenschaftlichen Medicin.* Frankfurt A. M. (Meidinger).
—— 1859. *Die Cellularpathologie in ihrer Begründung auf physiologischer und pathologischer Gewebelehre.* Berlin (Hirschwald).
VOGT, C., 1842. *Untersuchungen über die Entwicklungsgeschichte der Geburtshelferkrœte* (*Alytes obstetricans*). Solothurn (Jent & Gassmann).
WAGNER, R., 1839. *Lehrbuch der Physiologie für akademische Vorlesungen und mit besonderer Rücksicht auf das Bedürfniss der Aerzte.* Leipsig (Voss).
—— (edited by), 1842. *Handwörterbuch der Physiologie mit Rücksicht auf physiologische Pathologie.* Vol. 1. Braunschweig (Vieweg).
WELDON, W. F. R., 1898. Nature, **58**, 1.
WOLFF, C. F., 1759. *Theoria generationis.* Halae ad Salam Litteris Hendelianis.

The Cell-theory: a Restatement, History, and Critique

Part V. The Multiplication of Nuclei

By JOHN R. BAKER

(From the Cytological Laboratory, Dept. of Zoology, University Museum, Oxford)

Summary

1. The belief that nuclei arose by *exogeny*, without relation to pre-existent nuclei, was due mostly to Schleiden (1838). Kölliker (1843) supposed that new nuclei arose by *endogeny* within pre-existent nuclei.

2. Other early theories of the origin of nuclei contained a considerable element of truth. Many early workers thought that the ordinary nuclei of many-celled plants and animals multiplied by *division* (Bagge, 1841; Nägeli, 1844; von Baer, 1846), or by the *disappearance* of the old nucleus and its immediate *replacement* by two new ones (Nägeli, 1841; Reichert, 1846).

3. The history of the discovery of *mitosis* falls into three parts.

In the first (1842–70), chromosomes were seen accidentally from time to time, but no special attention was paid to them (? Nägeli, 1842; Reichert, 1847).

In the second (1871–8), metaphases and anaphases were repeatedly seen, placed in their right sequence, and recognized as normal stages in nuclear multiplication (Russow, 1872; Schneider, 1873; Bütschli, 1875; Strasburger, 1875).

In the third (1878 onwards), the main features of prophase and telophase were described and it was shown that the chromosomes replicated themselves by longitudinal division (Flemming, 1878–82). The separateness of the chromosomes in prophase and the constancy of their number were discovered (Rabl, 1885).

4. These researches proved that in ordinary mitosis the nucleus neither disappears completely nor divides. In certain Protozoa, mitotic division of the nucleus is a reality.

5. The indirect origin of cells, through the intermediacy of syncytia, was established by the work of Nägeli (1844), Rathke (1844), Kölliker (1844), and Leuckart (1858).

6. There is nearly always a cellular phase at some stage or other of the life-history of organisms, even when all the somatic tissues are syncytial. Certain Zygomycetes provide an exception.

Contents

[Quarterly Journal of Microscopical Science, Vol. 96, part 4, pp. 449-481, December 1955.]

Introduction

We are still concerned with Proposition III in the formulation of the cell-theory adopted in this series of papers: that is to say, with the proposition that *cells always arise, directly or indirectly, from pre-existent cells, usually by binary fission.* In Part IV of the series (1953) we traced the history of the discovery that cells multiply by division. As we saw in Part II (1949), cells are defined by their possession of *protoplasm* and a *nucleus.* In cell-division the protoplasm divides (Part IV, 1953). It remains to trace here the history of our knowledge of the way in which nuclei multiply. We shall concern ourselves with the discovery that nuclei are genetically related to pre-existent nuclei, and with the gradual revelation of the real nature of that relationship. The history of the discovery of the rest of the process of mitosis (the behaviour of the centrioles, asters, &c.) is irrelevant to our purpose and will not be considered. The remainder of the paper will be concerned with the indirect origin of cells from cells, by the production of syncytia and the subsequent formation of cells in or from these.

The discussion of Proposition III will be completed in Part VI of the series, which will deal with the continuity of cells from generation to generation.

A few words about the purpose of this series of papers would, I believe, be appreciated by some readers.

The cell-theory has been subjected to powerful attack. As a result, its validity has been questioned in zoological textbooks. I decided to study the evidence against it very carefully, in the original papers. Having done this, and examined the whole subject more widely, I reached the conclusion that the theory withstood the attacks. I then decided to try to persuade others of its validity. I found that I could only develop my argument and make myself understood by a historical treatment, with critical comments from the modern point of view. In many fields of science we must recognize an embryology of ideas: our modern outlook can only be fully grasped and assessed if we understand the causes that make us think as we do. This applies particularly to the cell-theory. Though I have great respect for the history of science, yet my main purpose in this series of papers has not been to write history, but to use a mainly historical method to establish what I believe to be an important truth about living organisms.

Early Theories of Nuclear Multiplication

Before telling the history of the discovery of cell-division, it was necessary, in Part IV of this series of papers, to describe the wholly erroneous theories that were for long entertained about the process by which cells multiply. Our understanding of the multiplication of nuclei has come in a different way. Some of the early theories were wrong, but they were not wholly wrong; and considerable interest attaches to them in so far as they led towards the discovery of mitosis. First, however, it is necessary to eliminate a theory that contained no element of truth. This was the theory that nuclei arise exogenously in what Schwann (1839, pp. 45 and 207–12) called a *Cytoblastem*, without any relation to pre-existing nuclei.

Exogeny

Valentin (1835, p. 194) appears to have been the first person to make a suggestion as to the origin of nuclei. He claimed that in the chorioid coat of the eye, nuclei arise by a process of precipitation. He confuses his remarks by calling the nuclei *Pigmentbläschen*, though they are colourless; the globules of pigment appear subsequently round them.

The theory that nuclei in general arise exogenously was due to Schleiden (1838, pp. 145–6). His ideas have already been given in detail in Part IV of this series of papers (1953, p. 416), and need only be briefly mentioned here. It may be remembered that in his view, a nucleolus appeared without any relation to a pre-existing nucleus, and the nucleus or *Cytoblast* was formed round this by deposition of a granular coagulum. This nucleus then produced a cell round itself. In his first paper (1838*a*), Schwann accepted Schleiden's scheme and applied it to animals. It was unfortunate that the first ideas about the multiplication of nuclei were completely wrong, yet supported by two famous investigators.

Henle (1841, pp. 153–4) was evidently affected by these beliefs. He shows a cartilage-cell (his plate V, fig. 6) with a nucleus containing a nucleolus at one end and a body resembling a nucleolus at the other. He suggests rather tentatively that a nucleus had just formed round one of the nucleoli. Kölliker at one time thought that nucleoli might appear spontaneously in certain cases, by the crystallization of granules in a homogeneous fluid, and that nuclei were subsequently formed round them (1844, pp. 143–4 and 150); but, as we shall see (p. 452), he supposed that nucleoli ordinarily multiplied by division within nuclei.

Nägeli at one time allowed that nuclei might originate without any relation to pre-existent nuclei (1846, see especially pp. 62–63).

During the eighteen-forties, the belief that new nuclei arose in some sort of connexion with pre-existing ones became quite general, but the older view still lingered on. One cannot fail to regret that Remak, who had done so much to elucidate the multiplication of cells, eventually retracted a little from the position he had taken up and began to equivocate. He came to believe that new nuclei might in certain cases originate independently of pre-existing ones. He thought that when small blood-vessels were developing in the cutis of the frog, new nuclei appeared that were not related to the embryonic nuclei; he remarked also that the stellate cells (presumably fibroblasts) of connective tissue developed without any known connexion with the cells of the embryo. He also thought that new cells originated in diseased tissues without any participation of pre-existing nuclei (1862, p. 282).

Remak was not the only distinguished investigator to continue to hold such views. Weismann (1863*a*), in his account of the development of the egg of

Chironomus, says that nuclei 'appear' (*erscheinen*) at the same moment over the whole of the blastoderm, which then separates itself off round each of them and thus forms uninucleate cells. Lankester (1875, pp. 38–41) thought that in the development of *Loligo*, the 'autoplasts' (nuclei of the yolk-epithelium) were of the same nature as the nuclei of the blastomeres, but for the most part of independent origin. (The actual origin of the yolk-epithelium of cephalopods was finally revealed in the next decade by Vialleton (1888).)

Endogeny

Kölliker appears to have been the only person who claimed that new nuclei arise endogenously within old. Mainly as a result of his studies of the embryology of nematodes and of the frog, he reached the conclusion that the nucleolus lengthens, constricts, and divides; a new nucleus then forms endogenously round each of the two nucleoli thus produced, within the membrane of the mother-nucleus (Kölliker, 1843; 1844, pp. 143–4, 150). All this he described in the puzzling nomenclature that has already been described (Part IV, p. 418). Later he summed up his opinion in very clear language. 'Nuclei and cells multiply by endogenous procreation', he wrote; 'nucleoli by division' (1845, p. 96).

The theory of endogeny bore little relation to the actual events of nuclear multiplication, but at least it involved a genetic relationship between old and new nuclei. The two theories to be discussed next came nearer to reality. Indeed, each of them revealed a considerable part of the truth.

Division

Even at the present day we often read in biological textbooks of nuclear division, though in fact, of course, typical nuclei do not, in any intelligible sense, divide. Exceptions to this are provided by certain Protozoa (see below, p. 474), and also by the polyenergid nuclei of certain other members of the same phylum (see Baker, 1948*a*); but the latter are better regarded as representing aggregations of many small nuclei. It is true that the nuclei of certain tissue-cells of higher animals have been supposed to multiply by 'amitosis'; but in fact it seems unlikely that such a rough-and-ready method could divide the gene-complex accurately enough to produce viable cells (though it might suffice in a syncytium). It is more likely that in these cases disguised mitoses occur, without regular, easily recognizable metaphases and anaphases.

Ehrenberg (1838) appears to have been the first to witness the multiplication of a typical nucleus. He saw clearly the nucleus of the protomonad flagellate *Monas vivipara*, but regarded it as the testis. He remarks (pp. 9–10) that it divides when the animal divides, and he gives a figure (his plate I, fig. IV*a*) professing to show a stage in this process.

Barry (1841, *a*, *b*, and *c*) thought that nuclei multiply by division, or rather fragmentation, and that the fragments become new *cells*. This view was accepted by Goodsir (1845, p. 2). Barry worked chiefly with the red blood-corpuscles of various vertebrates. His observations on this subject, however, are so unsatisfactory that they cannot be regarded as having contributed to knowledge. In an earlier paper (1839) there are some passages (on p. 361) that suggest at first glance that he witnessed nuclear multiplication in early mammalian embryos, but this is not so.

Bagge (1841) reported nuclear division in the early embryo of '*Ascaris nigrovenosa*' (= *Rhabdias bufonis*). He recognized that the duplication of the nucleus preceded that of the cell. His terminology is unfortunately misleading: he calls nuclei *cellulae* and cells *vitelli partes* or *globulae*. His illustration, here reproduced as fig. 1, is probably the earliest attempt to represent consecutive stages in the process of nuclear multiplication.

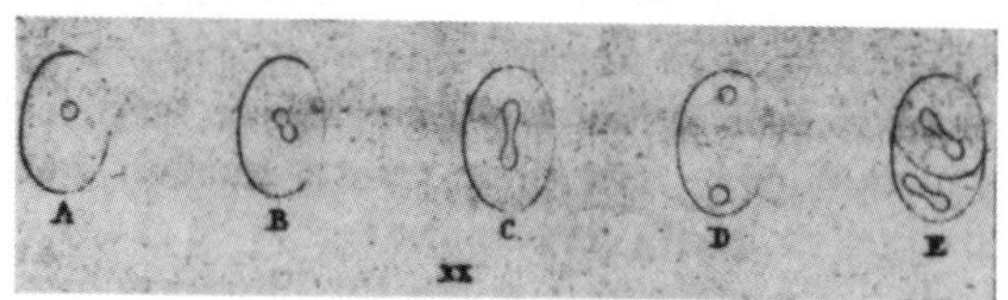

FIG. 1. Stages in supposed nuclear division in *Rhabdias bufonis*. Bagge, 1841 (fig. xx).

Remak (1841) studied blood-formation in the late chick-embryo. He noted that in dividing blood-cells the two nuclei were joined together by a stalk-like process (probably the remains of the spindle). Later (1845) he tried to follow the way in which nuclei multiply in the developing striated muscle of the frog tadpole. He was at first hesitant, remarking cautiously, 'I am not able to assert that the formation of new nuclei proceeds always from those already present, though several observations suggest this.' Further study of the same object convinced him that the new nuclei arise by division of the old (1855, p. 154), and he illustrates what he takes to be a division stage (his plate XI, fig. 5). Neither the text nor the figure gives any details of the process. Writing about nuclear multiplication in general, he admits (p. 174) that the process has not been elucidated with certainty, but claims that it clearly begins with a constriction. He leaves it undecided whether the nuclear membrane dissolves.

Valentin briefly described and figured a stage in the division or *Spaltung* of the nucleus of a cell in the membrane covering the auricles of the frog's heart (1842, p. 629 and plate VII, fig. 95 *bis*, *a* and *b*).

Breuer (1844, p. 31) and his associate, Günsburg (1848, pp. 361–2), claimed that in regenerating mammalian skin, nuclei multiplied by division or fragmentation (*sejunctione*). These authors may perhaps have been looking at the nuclei of polymorphs. Günsburg thought that the nucleus generally fell into as many pieces as there were nucleoli.

Nägeli (1844) described the division of the nucleus of a germinating spore of *Padina* (Phaeophyceae). His figure showing the two nuclei, supposed to have been produced by division of the old nucleus before the cell had divided, was reproduced on p. 433 of Part IV of this series of papers (1953). Nägeli subsequently came to regard division as the usual method of nuclear multiplication in plants (1846, pp. 68–69), though not, as we shall see (p. 457), the only method.

With von Baer we enter a new phase. His account of the process of nuclear multiplication was far fuller and more accurate than anything that had been published previously.

Perhaps because he published (in German) in a Russian journal, perhaps because his paper takes the shape of an informal, chatty letter from the seaside, von Baer's contribution (1846) to our understanding of the multiplication of nuclei has not received the credit it deserves. During his stay at Trieste he artificially fertilized the eggs of '*Echinus*' (*Paracentrotus*) *lividus* and watched the process of cleavage. In his description we can follow what was actually happening. It is very helpful to place beside his description a set of figures of the cleavage of the same animal made much later by Hertwig (1876). These figures are here reproduced in fig. 2, as an illustration of von Baer's paper. They show what can be seen in life. It will be noticed that Hertwig did not see the chromosomes (though he saw them clearly enough when he fixed and stained the embryo). Von Baer's great merit is that he gave a realistic description of what can be seen of mitosis when the chromosomes themselves are not seen.

Von Baer correctly identifies the nucleus (*Kern*) of the unfertilized egg. He tells us that on fertilization it sinks more deeply into the egg, and its limits become more difficult to see. He does not recognize the participation of the nucleus of the spermatozoon in the process, but thinks that the egg nucleus alone is the progenitor of those of the embryo. 'After some period of rest', he writes, 'this nucleus, up till now spherical, lengthens rather quickly by sprouting, as it were, at both sides; both ends swell, but the middle becomes thinner and soon divides completely, so that two comet-shaped nuclei with their tails lie opposite one another. Then, very quickly, the tail-shaped appendages pull themselves back into the spherical or vesicular masses, and one has two nuclei. . . . Before the division the original nucleus had already increased in volume; during the division this happens still more, so that each of the two new nuclei has apparently the size of the original one.' The egg now divides. 'Soon afterwards each of the two nuclei now begins to sprout out in the same way, and, dividing in the middle, changes into two new nuclei, round which the yolk-mass then likewise divides, and the whole egg resolves itself into four masses adhering to one another. . . . Quite similarly there follows the division of the quadrants, and indeed in such a way that the direction of the new sprouts stands at right angles to the immediately preceding ones. So it goes on with new divisions, for a nucleus forms itself in advance for each portion of the yolk by division of one that was produced earlier.' He remarks here that a pellicle is formed round the nucleus each time after a period of rest. 'Up to the division into 32 yolk-bodies, when the process is occurring quite regularly, I have been able to watch the division continuously.' The appearance is now for a time less clear. 'But still, when the embryo has left the egg-membranes and is moving itself by means of cilia, each granule or histogenetic element (vulgarly called a "cell") has a very evident nucleus, and they all appear to be derived from the original nucleus of the egg.' Movement of the larva now

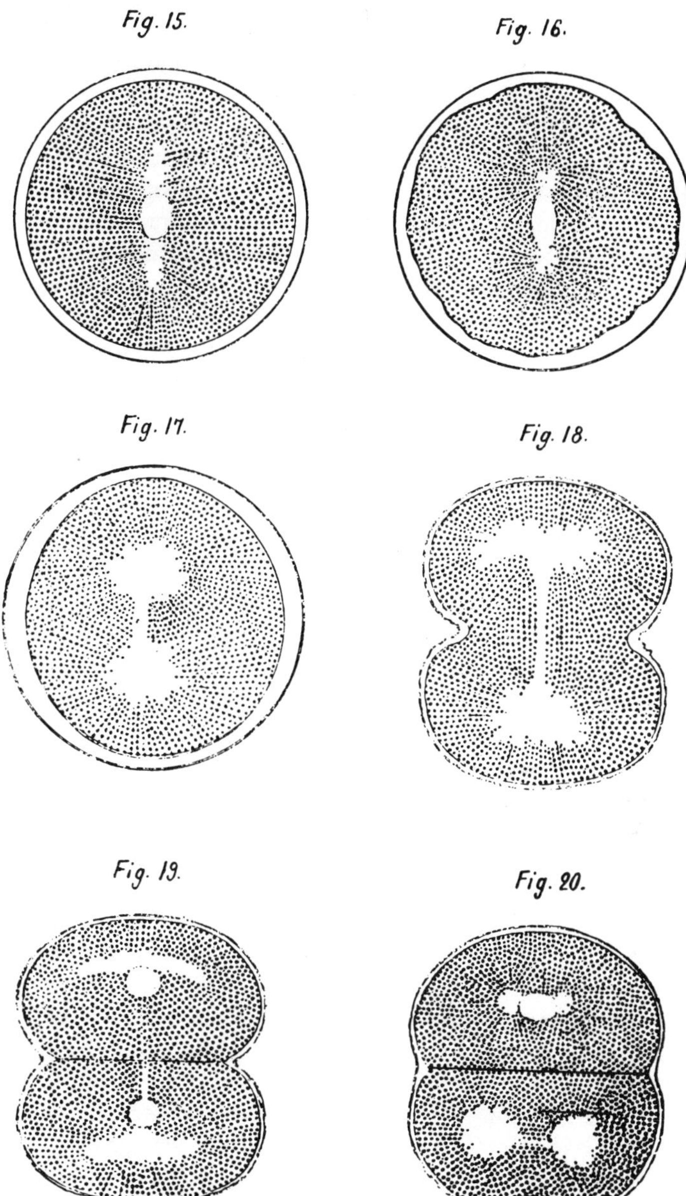

FIG. 2. Stages in nuclear replication in the embryo of *Paracentrotus lividus*, to illustrate von Baer's description. Hertwig, 1876 (plate XII).

makes observation difficult. 'But I have reason for the belief that the permanent tissue-constituents also arise from the original one by quite similar divisions. According to this, the divisions of the yolk would only be the beginnings of the histogenetic separation that progresses continuously up to the final formation of the animal' (1846, cols. 237–40).

Von Baer obviously saw the spindle and regarded it as a nucleus that had elongated in preparation for division. This outlook has not quite left us even today. One finds in textbooks statements to the effect that the spindle is formed from the nuclear sap. In fact, the nuclear membrane disappears during late prophase in most organisms other than certain Protozoa, and the nuclear sap then merges indistinguishably with the ground cytoplasm. Even if one disregards the fact that part of the spindle is often clearly formed in the cytoplasm while the nuclear membrane is still intact, it is still unjustifiable to derive the spindle exclusively from the nucleus. Even if this were not so, it would still be wrong to speak of nuclear multiplication by division; for in those cases in which the spindle-remnant survives to be divided across at cell-division, the products of its division are not incorporated in the new nuclei.

Virchow (1857) considered that nuclei ordinarily divide by a process of constriction. Gegenbaur (1858, pp. 9–10) saw something of nuclear multiplication in the cleavage of the egg of *Sagitta*. He admits that the details of the process escaped him, but he remarks that he saw a stage in which the nucleus was drawn out to a great length, and many were provided with constrictions. He presumably saw spindles and took them for elongated nuclei. He supposed that the actual division of the nucleus must take place very quickly.

Schultze (1861, p. 11) followed up his famous definition of a cell (see Part III of this series of papers (1952), p. 165) with a generalization on nuclear multiplication in the very next sentence. 'The nucleus and also the protoplasm', he wrote, 'are division-products of the same components of another cell.'

In his study of the development of *Musca vomitoria*, Weismann (1863*b*, p. 162) announced that each of the four pole-cells (primary germ-cells) divides into two, with simultaneous division of their nuclei.

As we shall see, a number of botanists had adopted the view that the nucleus disappears at cell-division and is somehow replaced by two new ones. Hanstein (1870) devoted a paper to the refutation of this belief. He worked chiefly with the parenchyma of various flowering plants. He satisfied himself that the nucleus did not disappear. He claimed that it was constricted by a delicate but optically perceptible halving-boundary (*Halbirungsgrenze*), and that when this process was complete, the two halves of the nucleus moved apart to opposite poles and a new cell-wall was formed between them (pp. 230–1).

Disappearance and replacement

The supporters of the theory of nuclear division performed a useful service by calling attention to the fact that two new nuclei are somehow derived from one old one, but they overlooked a rather obvious part of the usual process—the disappearance of the nuclear membrane and nucleolus. While some investigators were claiming that nuclei divided, others insisted that on the contrary a nucleus *disappears* and is *replaced* by two new ones. Each side in the controversy had seized upon one aspect of the truth.

If a nucleus completely disappeared and was then replaced by two new ones, the latter could be regarded as having arisen exogenously; but it seems desirable to draw a distinction between the origin of a new nucleus without any relation to a pre-existent nucleus, and the disappearance of one nucleus and its replacement by two new ones.

From his studies of pollen-formation, Nägeli (1841) concluded that when the mother-cell is about to divide, the cytoblast (nucleus) is absorbed. A new cytoblast then appears in each of two granular areas in the cytoplasm. Membranes form in such a way as to enclose each of the granular areas. The whole process is then repeated, with disappearance and replacement of the cytoblast. Thus four cells are formed, each with its nucleus. Alternatively, the four cells with their nuclei may be formed simultaneously after the disappearance of the nucleus of the mother-cell. Later, in a general account of pollen- and spore-formation, Nägeli repeats this general scheme, with the added complication that the original nucleus of the mother-cell, lying against the cell-wall, disappears and is replaced by a central nucleus, which in turn disappears and is replaced by four new ones or by two which are each subsequently replaced by two (1844, pp. 83–84 (p. 84 is accidentally numbered 48)).

As we have already seen (p. 453), Nägeli regarded division as the usual method by which the nuclei of plants multiply; but he retained his belief that in particular cases there is absorption and replacement (1846, p. 70).

Hofmeister thought it certain that the nucleus of the pollen mother-cell of *Tradescantia* underwent dissolution (*Auflösung*) and replacement by two new nuclei (1848*b*, col. 651).

Meanwhile, similar results were being obtained with animals. Reichert studied the egg and embryo of the nematode *Strongylus auricularis* (1846, pp. 201 and 255–6). He described the disappearance or *Hinschwinden* of the germinal vesicle and the mixture of its contents with the substance of the rest of the egg. A new nucleus was formed, but this again disappeared. A new one was formed in each of the first two blastomeres; these again disappeared before the next division. So the process went on. The nucleus underwent *Hinschwinden* before each division, the newly-formed cells contained no nucleus, and finally a new nucleus appeared in each. Reichert followed the repetition of this process up to the stage at which the form of the little worm had become visible. He illustrated his findings by careful drawings (his plate IX). Similarly, Krohn described the disappearance of the nucleus at each cleavage division in the ascidian *Phallusia*, and the reappearance of a nucleus in each newly-formed blastomere (1852, pp. 314–15).

Much later than this, at a time when chromosomes had often been seen, it was still supposed that the germinal vesicle of the primary oocyte did in fact wholly disappear when the polar bodies were about to be given off. This is perhaps not surprising in view of the large size of the vesicle in relation to that of the chromosomes. Thus van Beneden described the complete disappearance of the germinal vesicle in the rabbit (1875, p. 692). He remarks that at this stage the egg is what Haeckel called a *Cytode*; that is to say, a lump of cytoplasm not containing a nucleus (see Haeckel, 1866, pp. 273–4).

Auerbach also used the term *Cytode* for the cell in which the nucleus has disappeared before cell-division (1876, p. 258). He thought that the substance of the nucleus intermingled with the cytoplasm and dissolved in it. For this reason he termed the stage of mitosis at which the nucleus becomes no longer visible *die karyolytische Figur* (p. 222).

Mitosis

It is a fact that there is genetic continuity between old nuclei and new, but nuclei do not ordinarily multiply by division. It is a fact that the nuclear membrane and nucleolus disappear at cell-division, yet the whole of the nucleus does not vanish. The truth could only be established when the erroneous parts of each theory had been eliminated, and when the remainder had been integrated by the discovery of something important that had been overlooked by both—the chromosomes.

If we study the old papers in which the early descriptions of chromosomes appear, it seems at first almost impossible to give an intelligible exposition of the way in which our modern knowledge was achieved. Yet there is an evolutionary story to be told, for there were in fact *stages* in the history, more evident, no doubt, to us who look back than to those who lived through the events.

In the first stage there were mere accidental records of bodies that we can now recognize as chromosomes. In the next, metaphases and anaphases were repeatedly described, and came to be regarded as usual stages in the process of nuclear duplication. It was understood that the anaphase was subsequent to the metaphase. In the third and last stage, the prophase and telophase were carefully described, and the real nature of the genetic relationship between one nucleus and the two that succeed it was disclosed through the genius of Flemming.

The first period (1842–70)

In a general account of the changes of form of the nucleus, Henle (1841, pp. 193–4) makes some remarks that suggest strongly that he saw some of the stages of mitosis. He says that nuclei often become oval and then more elongated, and then change into thin striations. The nucleoli disappear and the nucleus then becomes decomposed into a row of little dots (*Pünktchen*). He mentions that nuclei are sometimes connected by threads. If he had left it at this, we should probably have believed that the striations represented the spindle and the *Pünktchen* the chromosomes; but he illustrates what he saw by drawings, and a study of these suggests that he was not looking at stages in cell-division after all.

Valentin (1842, pp. 630–1) says that after treatment with acids, nuclei are sometimes seen in the act of division, 'with granular accessory appendage'. One can only guess whether the granules were chromosomes. No details are given that could guide us.

Nägeli is the first person of whom we can say that he probably saw chromosomes. In his account of the formation of pollen in *Lilium tigrinum*, he describes how the cytoblast of the mother-cell vanishes and is replaced by a variable number of small cytoblasts, which are transitory and in their turn disappear (1842, pp. 11–12). He figures a cell containing seven of them (see

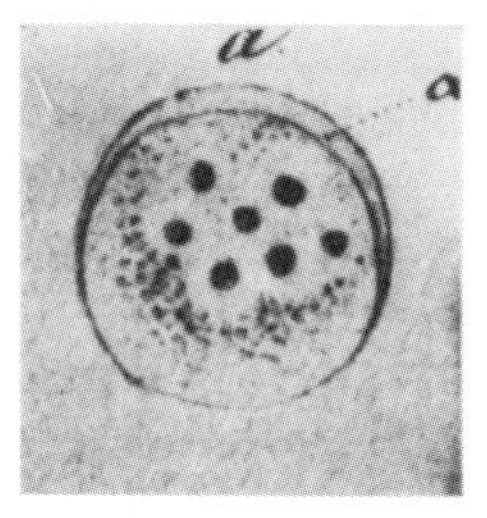

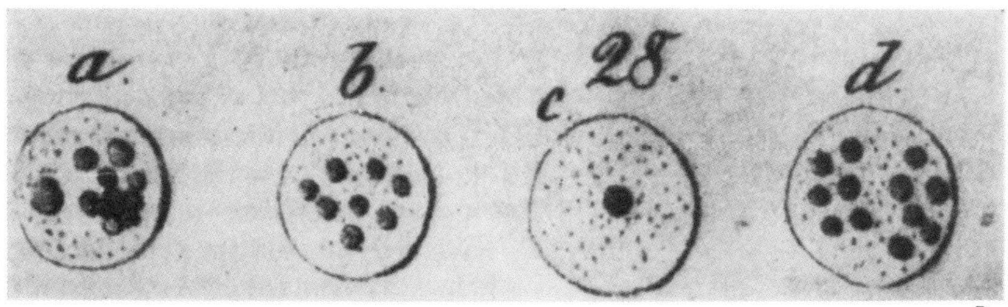

FIG. 3. These are probably the earliest illustrations of chromosomes. A, 'transitory cytoblasts' in the pollen mother-cell of *Lilium tigrinum*. B, the same, in pollen mother-cells of *Tradescantia*. Nägeli, 1842 (plates I, fig. 12*d*, and II, fig. 28; enlarged from the original figures).

fig. 3, A). When they have disappeared, the contents of the cell divide in two, and a new cytoblast is formed in each of the products of division. It is possible that these transitory cytoblasts were chromosomes, though doubt is engendered by the fact that in some cases there were only one or two of them in the cell. He saw similar appearances in the pollen mother-cell of *Tradescantia* (see fig. 3, B). It seems likely in this case that the bodies were actually chromosomes. Three of the cells illustrated contain 11, 8, and 12 such bodies. Nägeli thought it probable that when these transitory bodies had disappeared, a single, large cytoblast was formed (in the case of *Tradescantia*), and this divided in two (p. 13).

Later, in describing what was presumably the division of the pollen mother-cell of *Amaryllis*, he gives a figure of what he calls the *Kern* but is actually the spindle (1844, plate II, fig. 24*a*). He remarks that this body has 'dunkle körnige Anhänge'. These may have been anaphase chromosome-groups, but the figure is unfortunately on too small a scale to show them.

If Nägeli probably saw chromosomes, then Reichert (1847) certainly did so. He investigated carefully the spermatogenesis of *Strongylus auricularis*, noting that the developmental stages could be followed in time-sequence by passing along the testis from its blind end. After mentioning what were evidently the spermatogonia at the blind end of the tube, he goes on, 'Besides these, vesicles [cells] sometimes occur which contain two nuclei, of exactly the same microscopical constitution, but smaller [than those of the other cells], and also vesicles that are provided with no nuclei at all, but only with separate small granules (*Körnchen*); these vesicles cannot be mistaken for the clear vesicles that are perhaps of artificial origin' (p. 101). Serious artifact is indeed unlikely, for Reichert simply opened the testis at chosen places to let out the contents, and wetted the preparation with saliva or egg-white (p. 99). The *Körnchen* were evidently chromosomes, probably those of spermatogonial mitoses, for he thought that the nucleus actually disappeared altogether in what we should call the meiotic divisions (pp. 110–13).

Hofmeister (1848*a*) had the great merit of realizing that most contemporary cytologists were devoting a disproportionate amount of attention to the cell-wall and neglecting the nucleus. He decided to study nuclear phenomena in the pollen mother-cells of *Tradescantia virginica*. He found that as the cell grew, so did its nucleus and nucleolus; but eventually, at about the same time, the nucleolus and nuclear membrane disappeared. As we have already seen (p. 457), Hofmeister believed in the actual *Auflösung* of the nucleus in *Tradescantia*, and its replacement by two new ones. He thought, however, that when the nucleus had just dissolved, the albuminous material occupying what had been its site was in a particularly coagulable condition. He produced coagulation by the action of either water or tincture of iodine. Either of these agents produced separate *Klumpen* in the cell (cols. 427–30). These objects, which he supposed to be artificial coagulates, were in fact chromosomes (see fig. 4). He considered them to be of the same nature as the transitory cytoblasts of Nägeli.

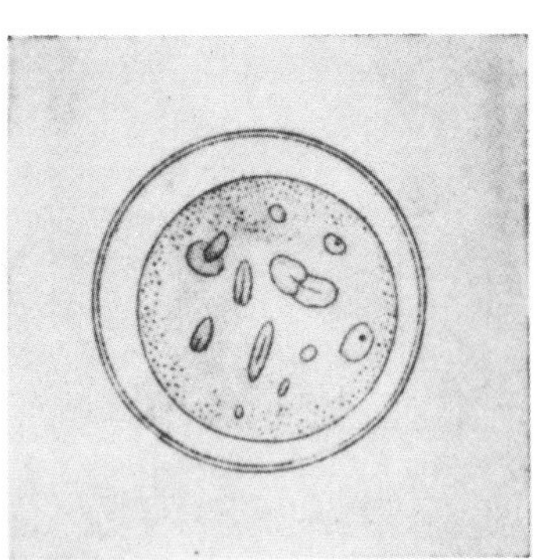

FIG. 4. The chromosomes in a pollen mother-cell of *Tradescantia virginica*, after treatment with tincture of iodine. Hofmeister, 1848 (plate IV, fig. 10*b*).

Nineteen years later Hofmeister still retained the same opinion. As a result of studies of the formation of pollen in various phanerogams and of spores in vascular cryptogams, he concluded that at the stage of disappearance of the nuclear membrane, the substance of the nucleus is easily coagulated as a little clot of strongly refractive substance, or else in the form of numerous, much smaller objects (1867, p. 81). The latter were undoubtedly chromosomes. He mentions that in spore-formation in *Equisetum* they are situated in the equator of the cell; in *Psilotum* they arrange themselves in a horizontal plate. Fig. 5 is a reproduction of Hofmeister's illustration of a meiotic metaphase in *Psilotum 'triquetum'* (= *nudum*). Fifty-four bodies, presumably chromosome-pairs, can be counted in this figure. The haploid chromosome number in this species is probably in fact 52 (see Manton, 1950, p. 239). Hofmeister's figure shows a remarkable resemblance to the same stage in *P. flaccidum*, as illustrated by Manton (her fig. 236; 52–54 chromosome-pairs). One cannot fail to be struck by such an accurate chromosome-count at this early date. Hofmeister once again attributed the appearance of chromosomes to artificial coagulation by water. In the circumstances it is surprising that he should have given such an accurate representation of their number.

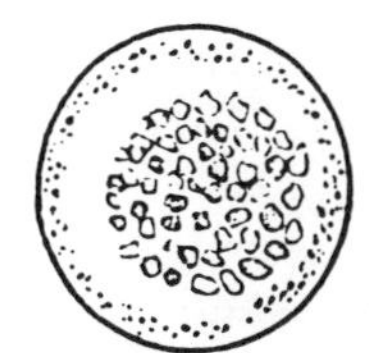

FIG. 5. The chromosomes in a spore mother-cell of *Psilotum nudum*. Hofmeister, 1867 (fig. 16, *e*).

Virchow (1857, p. 90) saw a cell with what he calls a branched nucleus among dividing cells in a cancerous lymph-gland. His figure suggests that he may possibly have seen a metaphase (his plate I, fig. 14).

Henle (1866) saw what were apparently pachytene stages (his figs. 263, 268) and perhaps the metaphase and telophase of the first maturation division (fig. 266) in mammalian spermatogenesis. His figures and descriptions are too vague, however, for certainty to be reached. He used acetic and chromic acids as fixatives (pp. 355–6), and one would expect the meiotic chromosomes to have been visible. Indeed, it is rather strange that there are so few early accounts of chromosomes in testes, for it would only have been necessary to examine the organs of almost any animal during the season of spermatogenesis to see at any rate the chromosomes of the first prophase. Such stages were seen much later by Spengel (1876) in the spermatogenesis of several genera of Gymnophiona. He compared them to Chinese writing (p. 31, see his plate II, fig. 33).

Chromosomes were probably seen by Krause (1870) in the epithelial cells of the surface of the cornea of various mammals. He called them granulated oval corpuscles (*Körperchen*). He thought that the nuclei of the epithelium multiplied by division, but did not claim that the corpuscles were necessarily connected with this process. Subsequently he gave an illustration of these bodies (1876, fig. 8, *f*). The figure seems to represent a late prophase.

The first period of chromosome studies, which started with Nägeli in 1842, ended with Krause's paper of 1870. Up to the latter date the descriptions and figures were vague and unsatisfactory, so that we generally cannot tell exactly what stage of mitosis or meiosis was seen, and we cannot even be certain, in some cases, that chromosomes were seen at all. No attempt had been made as yet to arrange the stages of mitosis in a time-series.

The second period (1871–8)

During the second period discoveries about chromosomes came with a rush. Kowalevski (1871) published the first figure of chromosomes that really resembles the object. One can tell instantly on looking at his drawing that he saw an anaphase (see fig. 6). His object was a section through the embryo of the lumbriculid worm '*Euaxes*' (= *Rhynchelmis*), at the moment when the first set of micromeres was being given off. He used chromic acid for hardening the egg. He calls the two groups of chromosomes *zwei körnige Anhäufungen* (p. 13). He regarded them as representing division-products of the nucleolus.

Russow (1872) carried our knowledge of chromosomes much further in a study of spore-formation in vascular cryptogams. He found that at the division of the spore mother-cell, a *Stäbchenplatte* was formed (pp. 89, 126). He saw these metaphase plates particularly clearly in ferns and Equisetales. Leaving for a moment the cryptogams that form the main subject of his very long

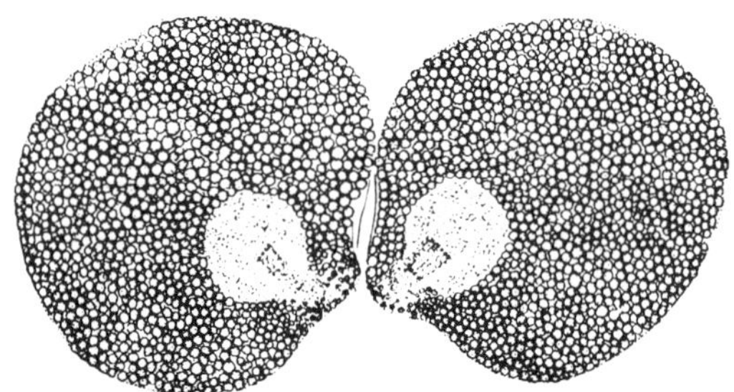

Fig. 6. Anaphase at the formation of the first quartette of micromeres in *Rhynchelmis*. Kowalevski, 1871 (plate IV, fig. 24).

paper, he remarks that he has seen these plates most clearly of all in the pollen mother-cells of *Lilium bulbiferum*. He then proceeds to the first serious attempt ever made at a description of chromosomes. He remarks (p. 90) that in *Lilium* they are short, worm-shaped corpuscles or slightly curved rodlets, colourless, pale, and faintly refractive, scarcely detectably stained by iodine, and almost instantaneously dissolved by alkalis (even at great dilution) and by ammonium carminate; also by chlor-zinc-iodide, without becoming coloured. He noted the same chemical behaviour in the chromosomes of the vascular cryptogams. He distinguished clearly between the *Stäbchenplatte* and the subsequently-formed *Körnerplatte* (cell-plate).

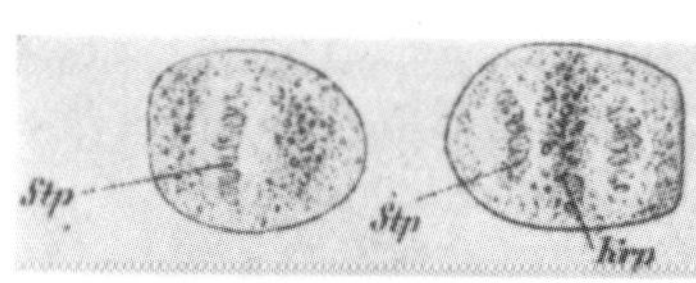

Fig. 7. Metaphase and anaphase in the spore mother-cell of *Ophioglossum vulgatum*. Russow, 1872 (plate VI, figs. 121 and 122).

Russow also saw anaphase chromosome-groups (see fig. 7). He called each of them a secondary *Stäbchenplatte*. He did not explain how they arose, and evidently thought they were the metaphase plates of the next division (pp. 127, 204). He noted that when there is a *Stäbchenplatte*, there is never a nucleus, and put forward the possibility that the plate is formed from the nucleus. He noted (p. 90) that when a secondary plate is viewed from the side, it resembles a very granular nucleus. This suggests that he saw a telophase.

Russow strongly denied that the chromosome plate is an artifact, as Hofmeister had supposed. He saw it in the intact sporangia of *Polypodium* and '*Aspidium*' (= *Dryopteris*).

Schneider's paper of 1873 constitutes a landmark in the history of our knowledge of chromosomes. He followed carefully the cleavage of the egg of the rhabdocoel *Mesostomum ehrenbergii*. By using acetic acid as fixative, he clearly saw the *dicke Stränge* or chromosomes, and noticed that one half of them went to one pole and the other half to the other (see fig. 8). He saw mitosis not only during cleavage, but also during later embryonic stages and in the adult (pp. 113–16 and plate V, fig. 11). He knew that the nucleus had been thought by others to disappear at cell-division, but he considered it probable that the process he had described in *Mesostomum* actually occurred in these cases. It is evident that he thought that amitotic division also took place. Schneider's paper is above all important for its clear seriation of metaphase and anaphase.

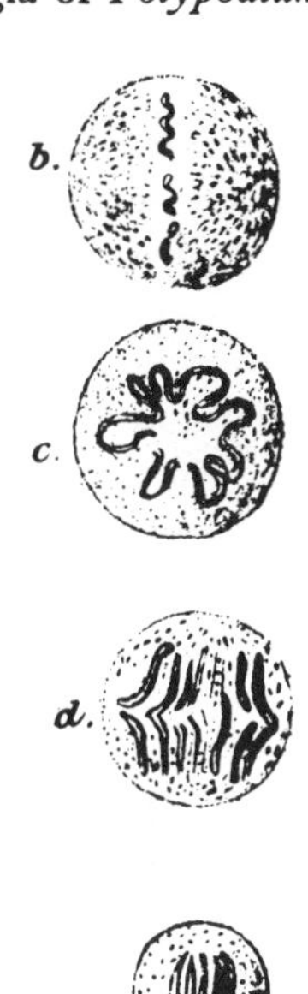

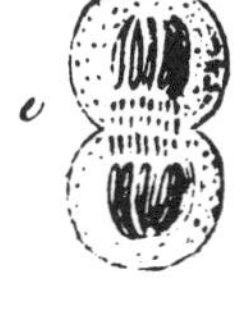

FIG. 8. The earliest figure showing stages in mitosis in correct sequence. The first cleavage of the egg of *Mesostomum ehrenbergii*. Schneider, 1873 (plate V, fig. 5).

Tschistiakoff (1875, *a* and *b*) saw metaphases and illustrated them clearly in spore mother-cells of *Isoetes* (Lycopodinae) and various ferns, and also in pollen mother-cells of *Magnolia* (his plate I, figs. xx and xxiv). He called the chromosome-plate a *Körnchenlamelle* (1875*a*, col. 1). He made the serious mistake of supposing that the two new nuclei were formed at the poles of the spindle while the chromosomes were still on the metaphase plate (1875*b*).

A paper of this period by Ewetsky (1875) is remarkable because it contains the first reasonably good figure of a prophase (see fig. 9). The cell is from the endothelium of Descemet's membrane in the eye, in regeneration following operational damage. Like Russow, Ewetsky called the chromosomes vermiform structures. Like so many others, he thought that the spindle was the nucleus and that it divided, merely enclosing the anaphase chromosome-groups at its ends.

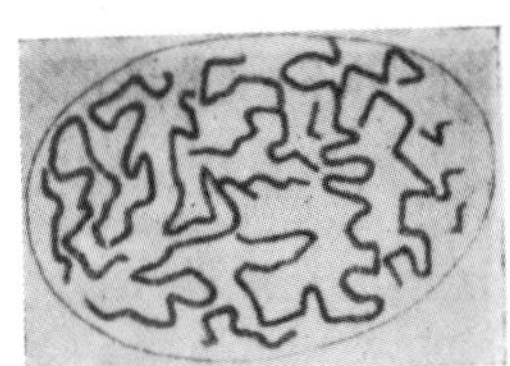

FIG. 9. An early representation of prophase. An endothelial cell of Descemet's membrane. Ewetsky, 1875 (plate V, fig. 7).

Bütschli (1875) studied polar body formation and cleavage in the nematode *Cucullanus elegans*, a parasite of fresh-water fishes. He saw metaphases and anaphases distinctly. Unfortunately his paper is not illustrated. The drawing shown here (fig. 10) is from his publication of the following year. He gives a particularly clear account of the occurrences at the first cleavage (1875, pp. 211–12). He says that the nucleus becomes invisible and a longitudinally striated, spindle-shaped body appears. Each fibre is swollen at the equator of the spindle into a *Korn* or *Körnchen*, and if one looks from the end of the spindle one sees a ring of granules (*Körnchenkreis*). The error of supposing that chromosomes were swollen regions of spindle-fibres was hard to eradicate in subsequent years. It may be remarked that Bütschli's name for the group of chromosomes at metaphase was a much more realistic one than the term

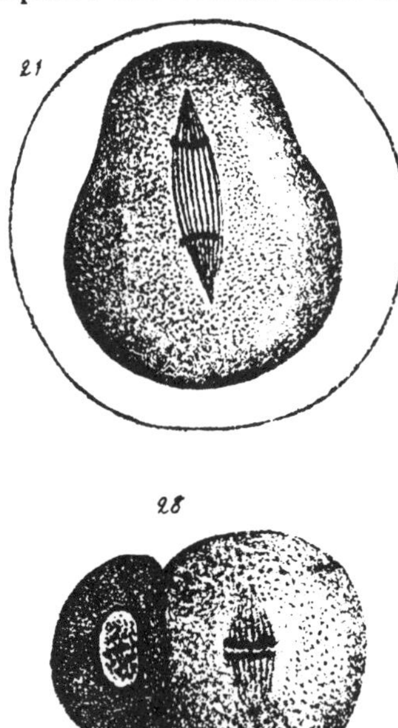

FIG. 10. Typical illustrations of mitosis by Bütschli. Anaphase of the first cleavage and metaphase of the second in *Cucullanus elegans*. Bütschli, 1876 (plate III, figs. 21 and 28).

Platte, which still survives; for there is no real resemblance to a plate unless the chromosomes happen to be very small, very numerous, and very close together. From the single ring of granules now arise two rings, and these move apart. Bütschli uses the expressive term *Auseinanderrücken der Körnchenkreise* for the anaphase. Although he used acetic acid as a fixative in the investigation, it seems possible that he watched the anaphase in life. He noticed that the spindle and chromosomes disappeared and a new nucleus was formed at each pole. He gives no details of these occurrences. He showed the negative merit of not regarding the new nuclei as division-products of the spindle.

Bütschli, like others, had already (1873) seen and figured spindles (but not chromosomes) in the micronuclear mitoses of *Paramecium*, and he recognized the same object in *Cucullanus*. In 1873 he had regarded the spindles as seminal vesicles.

In his paper of 1875 Bütschli described the fusion of chromosomal vesicles to form a nucleus, but he did not relate the vesicles to chromosomes. Such vesicles had already been reported by Remak (1855, p. 139 and plate IX, fig. 14), Lang (1872), and Oellacher (1872, pp. 410–11 and plate XXXIII, figs. 29–36). These investigations, however, did not help towards the elucidation of mitosis.

Strasburger was drawn into the study of mitosis by his interest in the process of fertilization in conifers. He worked particularly with the spruce, *Picea vulgaris*. The first two mitoses after fertilization escaped him, but he studied the multiplication of the four resulting nuclei, which are situated at the end

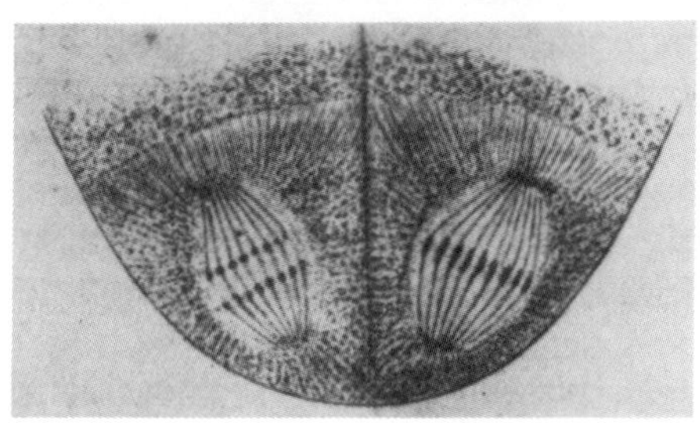

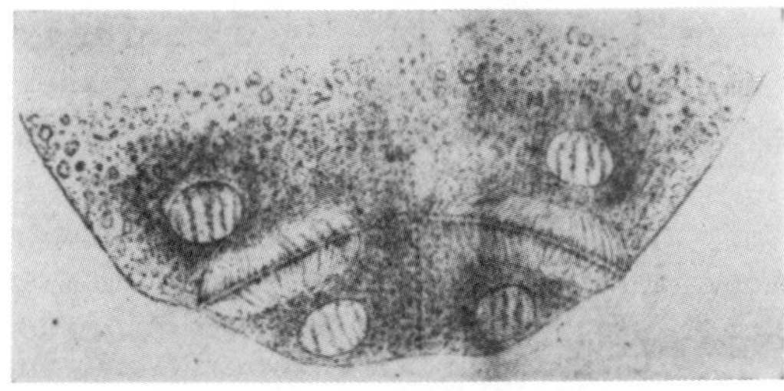

FIG. 11. Typical illustrations of mitosis by Strasburger. Metaphase, anaphase, and telophase in the multiplication of the nuclei at the lower end of the ovum of *Picea vulgaris*. Strasburger, 1875 (plate II, figs. 27 and 30).

of the ovum that is farthest from the micropyle. He used material fixed in alcohol, without staining. He saw metaphases and anaphases clearly (see fig. 11, upper figure). His description (1875, pp. 26–27) shows that he thought the process was one of actual nuclear division. He regarded the spindle, with its equatorial *Platte* of *Stäbchen* (chromosomes), as the nucleus. When the *Platte* separated into two, at the beginning of anaphase, the nucleus had divided. Each daughter-nucleus was formed by the fusion of the chromosomes of one plate with one another and with half the spindle. Strasburger made scarcely any attempt at a description of prophase or telophase, though he did illustrate a telophase (fig. 11, lower figure).

Strasburger realized the necessity of studying the process in life, so as to be sure that alcohol did not produce artificial appearances and also so as to be able to place the stages in their proper sequence with certainty. He found suitable material in *Spirogyra*. He followed the whole process of mitosis in the living alga (1875, pp. 33–46 and plate III), though he was not able to see much of prophase or telophase beyond the disappearance and reappearance of the nucleolus. He also studied alcohol-preparations, and confirmed their reliability. (Several years later (1879) he found a particularly suitable object for vital studies of mitosis in the staminal hairs of *Tradescantia*, immersed in 1% cane-sugar solution.)

Having found close resemblance between the processes of mitosis in such widely different plants, Strasburger now investigated it in many other diverse forms, and saw similar metaphases and anaphases over and over again. The cells he studied all had short chromosomes, and it is probable that this fact prevented a more complete understanding of what really happens in mitosis.

Strasburger was not content to study plants only. He knew of Bütschli's work on *Cucullanus*, and got into touch with him. Bütschli provided him with unpublished figures of mitosis in *Cucullanus* and of meiosis in *Blatta*, and Strasburger recognized the similarity to what he had seen in plants. Strasburger himself studied mitosis in mammalian cartilage (1875, pp. 186–9) and especially in the cleavage of the ascidian, *Phallusia mamillata* (pp. 189–97).

Strasburger brought the whole of his work together in his justly famous book, *Ueber Zellbildung und Zelltheilung* (1875), in which he included (with full acknowledgement) some of Bütschli's unpublished figures. This book was by far the most complete account of cell-division available at the time, and served to show the universality of mitosis as the ordinary process of nuclear multiplication.

At the time, the writings of Bütschli and Strasburger attracted far more attention than those of the Russian cytologists and Schneider. A number of authors were quick to recognize, in their own research-material, the descriptions given by the two former workers. Van Beneden was one of the first to come under their spell (1875). He studied the process of nuclear duplication in the ectoderm of the rabbit embryo, after fixation with osmium tetroxide and staining generally with picrocarmine (a favourite combination with early students of chromosomes). He recognized the separation of the equatorial plate of refringent globules or *bâtonnets* into two *disques nucléaires*, and the movement of these apart from one another at anaphase. He noted correctly that the new nuclei were formed from the disks, which swelled up at the expense of the surrounding cytoplasm and became less and less easily stainable. Later, in the course of the work that resulted in the foundation of the group Mesozoa, van Beneden saw mitosis in the cleavage of the cell that gives rise to the infusiform embryo of *Dicyemella* (1876, pp. 48–52; plate I, fig. 28; plate III, figs. 2, 4, 11).

Mayzel (1875), also influenced by Bütschli and Strasburger, saw various stages of mitosis in the regenerating corneal epithelium of the frog, including a metaphase with radially-arranged chromosomes (p. 851).

The rarity of Strasburger's first edition reflects the publisher's underestimation of the interest of this new line of research. A new edition was quickly produced (1876), with advice from Bütschli in correspondence and conversation. In the same year Bütschli produced an immense paper (1876), profusely illustrated with figures of mitosis in the cleavage of a leech, of *Cucullanus* (see fig. 10), and of *Limnaea*, of meiosis in the testis of *Blatta*, and of nuclear changes in the conjugation of various ciliates. It is strange to note his pre-occupation with side-views of metaphases and anaphases. Prophases he scarcely noted. He shows one, however, in *Cucullanus* (his plate III, fig. 20) and suggests that it may represent a preliminary stage (p. 226). If only he had studied metaphases more often in polar view, he might have made important discoveries about the constancy of chromosome number. He still describes the equatorial plate as consisting of the thickened parts of the spindle-fibres (p. 219). Telophase still eludes him, but he thinks that new nuclei must arise from the groups of chromosomes at the two ends of the spindle (p. 220).

Balbiani (1876) found an excellent source of mitotic figures in the ovariolar epithelium of the nymph of the grasshopper *Stenobothrus*. He saw the prophase chromosomes as short, unequal rods and followed them through all the stages of mitosis till they fused at telophase to form a mass that became vacuolated; a membrane then appeared round it. He says that each of the *bâtonnets* cuts itself into two halves, but he gives no indication of the direction of the cutting. This short paper is, for its period, a remarkably complete account of mitosis in a single kind of cell.

The study of mitosis was now taken up actively by O. Hertwig, who produced a succession of papers on polar-body formation and cleavage in leeches, heteropods, echinoderms, and frogs (1876, 1877, 1878, *a* and *b*). An exponent of the osmium / carmine technique (and also of others), his careful studies were marred only by a tendency to follow Bütschli in regarding the chromosomes as swellings of the spindle-fibres. These papers were important because they revealed new facts about polar bodies and fertilization rather than because they established new details of the process of mitosis; but they helped to show how widely applicable were the findings of Bütschli, Strasburger, and the rest.

Eberth (1876) saw mitotic figures in the regenerating cornea of the frog and rabbit, and compared them with Strasburger's descriptions. The latter, in an interesting critique of Eberth's findings, discusses the origin of the spindle (1877, p. 522). He denies that it is derived simply from the nucleus. Sometimes there is no distinction between nuclear sap and cytoplasm at the time when the spindle is formed: the two have become continuous with one another. This message from the past deserves attention at the present day.

One cannot better comprehend the deficiencies of knowledge about mitosis at the close of the second period than by studying the third edition of Strasburger's book, which was published a little later (1880). Though the stages of metaphase and anaphase were by this time so familiar, they were not in the least understood. Strasburger still believed that the *Kernplatte* or equatorial 'plate' of chromosomes became divided at metaphase, and that this division was of a hit-or-miss nature (pp. 331–3). He thought that rod-shaped chromosomes ordinarily arranged themselves along the length of the spindle. Division took place in the same way whether the chromosomes were rods or granules.

If any part of a rod or granule was in the equatorial plane at metaphase, that granule or rod was divided across at that place. Those granules or rods that lay nearer one pole of the spindle passed towards that pole without division. Longitudinal division never occurred except in those particular cases in which long chromosomes arranged themselves at metaphase wholly in the equatorial plane. The splitting then occurred at metaphase.

The third period (1878 onwards)

Five years after the publication of Strasburger's third edition the chief facts of mitosis had been established, chiefly by the brilliant researches of Flemming and Rabl.

Strasburger, as we have seen, had already followed mitosis in the living cells of *Spirogyra*. Particular stages in mitosis had been seen in the living cells of animals. Mayzel (1877), for instance, had put various cells of vertebrates in aqueous humor and seen the stages that he had previously (1875) studied in fixed preparations. No one, however, had watched the actual succession of the stages in animal cells. Now, in 1878, three papers were published by investigators who had seen the process of mitosis unroll before their eyes. Schleicher was the first in the field, with a very short account of mitosis in living cartilage-cells of various Amphibia (8 June). Peremeschko, who had read Schleicher's paper, was next (27 July). He studied epithelial cells, fibroblasts, leucocytes, and endothelial cells of blood-vessels in the tails of young newts ('*Triton*' (*Triturus*) *cristatus*). Schleicher's and Peremeschko's studies were of importance in confirming beyond question the order of the most striking events of mitosis. Peremeschko remarked that the threads were sometimes *knäuelförmig* at the beginning of the process; so evidently he saw something of the prophase. Neither of these authors, however, added any important new facts.

Flemming had already (1877) chosen the salamander (*Salamandra maculata*) as his cytological research-material, on account of the large size of the cells and nuclei in this animal. He now started his research on cell-division, and read a paper on the subject at Kiel on the very day (1 August) on which he first saw a copy of Peremeschko's. This paper (1878) was short, in marked contrast to his massive later contributions. It contains more than just the foreshadowings of the important discoveries that were to come. His research-material was again the salamander, especially its larva. He studied the urinary bladder, the epithelium of the skin and gills, cartilage, connective tissue, the endothelium of blood-vessels, and blood-cells. He followed the whole process of mitosis in life and also studied fixed and stained preparations. Flemming was outspoken in his criticism of Strasburger's scheme of mitosis. Here in this paper one finds the first serious attempt at a description of the *Anfangsphasen* or early prophase-stages. Flemming definitely derives the chromosomes (*Fäden*) from the stainable substance visible in the form of a *Gerüst* in the interphase nucleus. He traces the gradual shortening and thickening of the trabeculae of the *Gerüst* to form finally the chromosomes of the metaphase *Stern*. He considers that the disappearing nucleolus supplies material to the thickening chromosomes. He notes that the chromosomes are split longitudinally throughout their length in (late) prophase, a discovery of the first importance, but in this paper he does not trace one longitudinal half to each pole in anaphase. He gives no exact account of what happens at metaphase. He sees the chromosomes move apart at anaphase, and describes the changes of telophase as resembling those of prophase in reverse. Thus for the first time the chromatin was followed through from one resting nucleus to the next. As a result of his studies, Flemming denied that the nucleus could be said without qualification to divide.

Flemming's choice of organisms with long chromosomes as his research-material, both in this early work and later, undoubtedly helped him to elucidate the main features of mitosis.

Flemming now began to produce a succession of papers, the main purpose of which was to show the uniformity of the process of mitosis. He denied that amitosis had been proved to exist in the tissues of animals, except possibly in leucocytes. Indirect nuclear division, with *Fadenmetamorphosen des Kerns*, was the rule (1879*a*, pp. 21–22). He now reverted to the *Längsspaltung* of the chromosomes. He saw the longitudinal split in both prophase and metaphase (see fig. 12) and suggested tentatively that one longitudinal half of each thread might go to each daughter-nucleus (1879*b*, p. 384). In the same paper he describes the prophase. He makes the mistake of supposing that the chromosomes are joined end to end into a continuous *Knäuel*. This he describes as a thin thread, which thickens and eventually breaks across into separate chromosomes. He once more describes the telophase as prophase in reverse, and gives a tabular synopsis of the whole process, arranged to bring out this resemblance (p. 392). He follows the telophase *Knäuel* into the network (*Gerüst*) of the interphase nucleus.

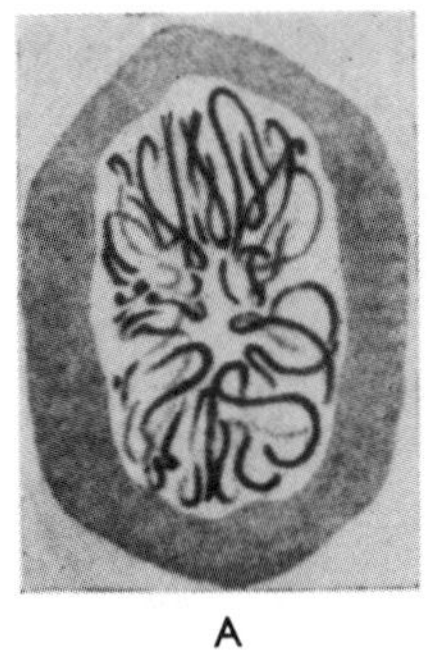

A

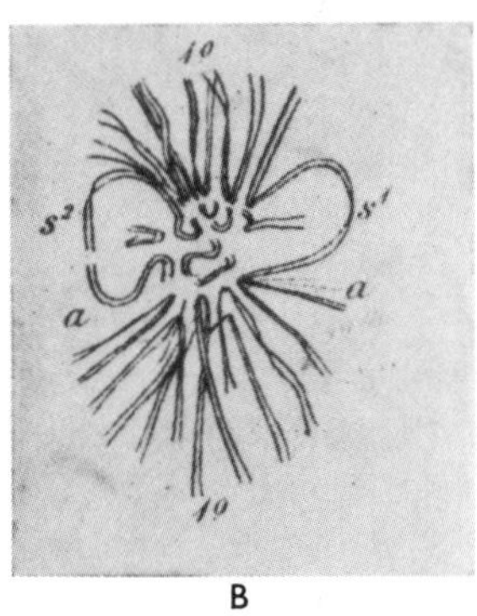

B

FIG. 12. The earliest figures showing the longitudinal splitting of chromosomes. A, epithelial cell from the gill of a salamander larva, to show the longitudinal split in some of the chromosomes at prophase. B, endothelial cell in a small blood-vessel of a salamander larva, to show the longitudinal split at metaphase. Flemming, 1879*b* (plate XVII, figs. 7 and 11).

Flemming emphasized that longitudinal splitting occurs constantly in diverse kinds of cells (1880, p. 212). He doubted Strasburger's belief that there were considerable differences between one case and another, and turned to plant material to find whether his scheme applied there also. Removing the coverslips from someone else's preparations of the embryo-sack of *Lilium croceum*, he restained them to his own satisfaction and was able to confirm that what he had so often observed in animals occurred also in this case (1882*a*, p. 43). He denied that the chromosomes fuse at metaphase to form a plate.

Flemming's great achievement was his discovery that in mitosis one longi-

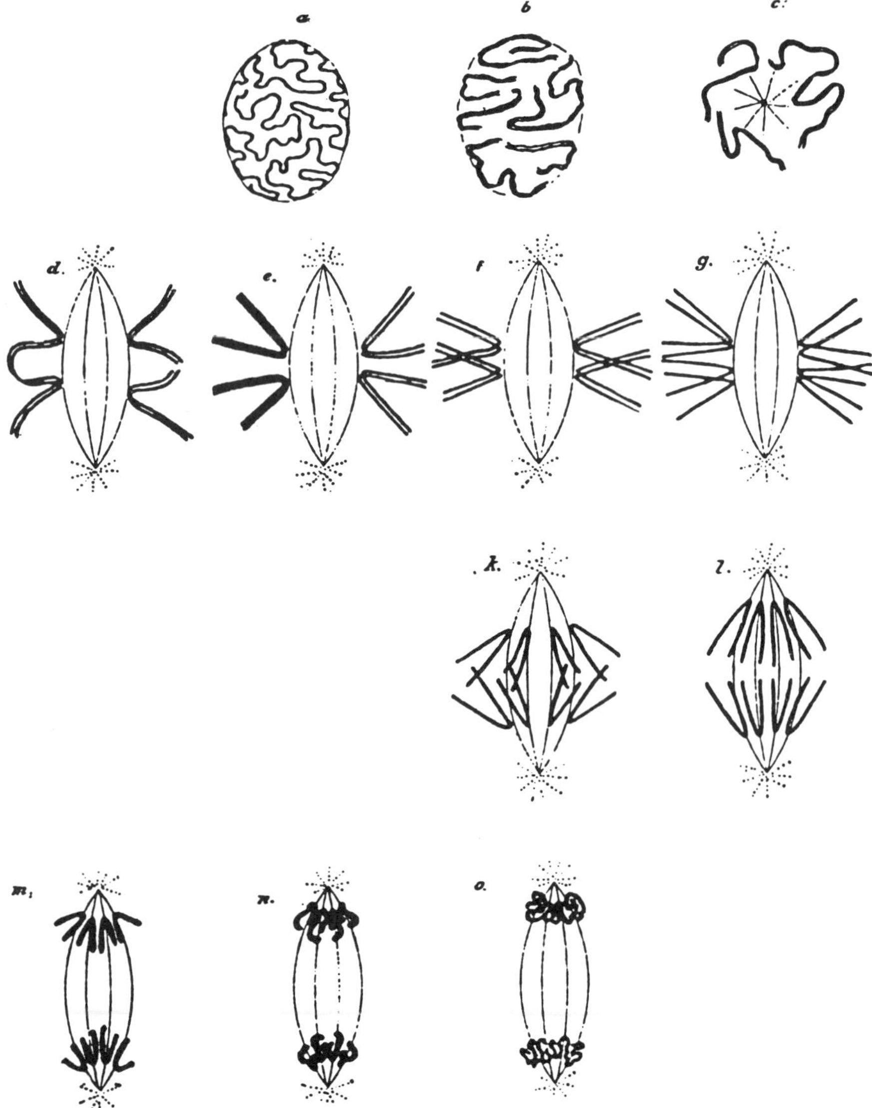

FIG. 13. Diagram of mitosis. Flemming, 1882*b* (plate VIII, fig. 1). (*h* and *i* are here omitted. Flemming used them to illustrate Strasburger's opinions.)

tudinal half of each chromosome goes in each direction, so that each daughter-nucleus is formed from a complete set of longitudinal halves. He brought together his ideas on the basic plan of mitosis in a diagram published in his book, *Zellsubstanz, Kern und Zelltheilung* (1882*b*). The diagram is reproduced here as fig. 13. The only error of any importance is the joining together of the chromosomes in prophase, end to end, to form a continuous *Knäuel*. The achievement represented by this diagram can be appreciated when it is remembered that only a few years earlier, almost nothing was known of prophase or telophase, and the most essential fact of mitosis—the separation of the chromosomes into two groups of longitudinal halves—was quite unknown.

Rabl's work (1885) was complementary to Flemming's, for it corrected the latter's main error and made good his main deficiency. His material was the epidermis of the floor of the mouth and of the gills of the larva of *Salamandra maculata*, and various organs of *Proteus*. He showed that in the prophase, the chromosomes are not joined end-to-end in a continuous *Knäuel*, but are separate from one another from the first. Further, their number is the same as at metaphase, and this number is always the same (24) in various cells of the two species studied. Rabl carried out his extremely laborious work with the utmost care and skill. One of his figures of a prophase, with separate, numbered chromosomes, is shown here in fig. 14. By 1880, as Boveri (1904, p. 4) remarked, the investigations of Flemming, Strasburger, van Beneden, and others were already leading to the conclusion that the number of chromosomes in the cells of each species was the same or nearly the same, but the paper that decided the issue was Rabl's of 1885. Boveri had great admiration for the Austrian cytologist. He writes of his 'wonderful perseverance and observational capacity' (1888, p. 5). Rabl studied carefully the early prophase and late telophase stages, and his findings in this field, taken in conjunction with his establishment of the constancy of chromosome number, formed the basis on which Boveri started to build his theory of the individuality or continuity of the chromosomes.

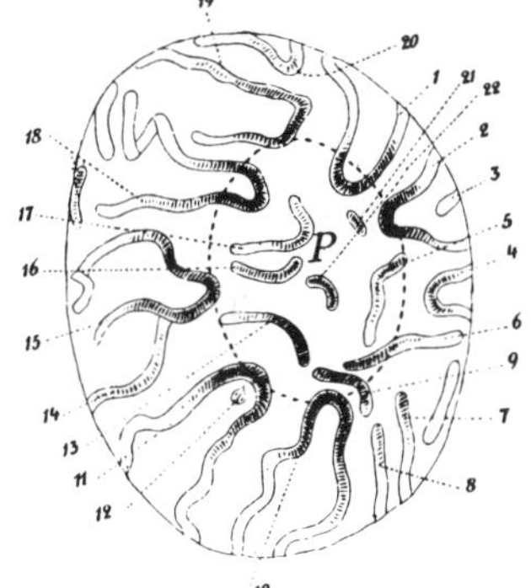

FIG. 14. The chromosomes in a prophase nucleus of an epidermal cell of the larva of *Salamandra maculata*. Twenty-two of the twenty-four chromosomes are shown. Rabl, 1885 (plate XIII, 6th figure).

Nomenclature

It is curious that cytologists should have been so slow to suggest international technical terms for the description of mitosis.

Schleicher pointed out the inconvenience of the multiplicity of terms used for the process itself, and suggested *Karyokinesis* (1879, p. 261). Flemming remarked that even in direct nuclear division, there is movement in or at the nucleus, which is what Schleicher's term means. He therefore suggested the substitution of *Karyomitosis* for it, to indicate 'thread-metamorphosis in the nucleus'. Instead of having to say 'nuclear division figures', one might use the short word *Mitosen* (1882*b*, p. 376). It must be regretted that Flemming was so far influenced by his profound study of the long chromosomes of the salamander as to choose this word, for in very many organisms the metaphase chromosomes cannot by any stretch of the imagination be described as threads.

The terms *Prophase*, *Metaphase*, and *Anaphase* were coined by Strasburger (1884, pp. 250 and 260). He did not use the terms metaphase and anaphase exactly as they are used today. For him, the metaphase continued until the daughter-chromosomes were quite separate from one another. Today, the anaphase is usually regarded as starting as soon as the spindle-attachments begin suddenly to move towards the poles. There would be little or no difference between the two usages of the word, however, in the case of the short chromosomes with which Strasburger usually worked. His anaphase included the modern anaphase (or late anaphase) and also telophase. His failure to give a separate name to the latter reflects his lack of attention to the final stage of mitosis.

The word *Telophase* was coined by M. Heidenhain ten years later (1894, p. 524). Curiously enough, he did not define it by any change in the chromosomes. His telophase started directly the centrosomes (*Mikrocentren*) left their positions at the poles of the spindle to migrate to their final sites in the new cells.

Interphase was introduced by Lundegårdh (1913, p. 211) to indicate the period between two closely consecutive mitoses. He drew a distinction between interphase and the *Ruhezustand* that follows mitosis when another division will not occur, or will be indefinitely delayed (pp. 213–19). He described differences between nuclei in interphase and those in the resting state.

Very diverse terms were used for chromosomes before that word was at long last introduced. They were called *Fäden*, *Kernfäden*, *Schleifen*, *Stäbchen*, *stäbchenförmige Körner*, *Körnchen*, and *chromatische Elemente*. Waldeyer certainly performed a useful service when he introduced *Chromosomen* (1888, p. 27). It would have been difficult to choose a shorter word so applicable to the object named in all its variant forms.

Comment

It is a strange fact that some of the best early workers on chromosomes continued to believe that nuclei ordinarily multiplied by division. Auerbach did not fall into this error. He contributed to the subject a short paper that did not receive the attention it deserved. As we have seen (p. 458), he at first believed in the actual solution of the whole of the nucleus at cell-division. Later, when the participation of chromosomes in the process had been repeatedly described, he justly claimed that there had been an element of truth in his belief. He denied that the spindle is derived solely from the nucleus, and that the main mass of it participates in the formation of the new nuclei; and he protested against the statement that nuclei divide (1876).

To resolve fully the question whether mitosis is nuclear division it would have been necessary to know the structure of the interphase nucleus, but even today this is a subject on which we are very imperfectly informed. Very large chromosomes, especially those of certain Liliaceae and Urodela, have attracted a great deal of attention from students of mitosis, but there is reason to believe that the nuclei to which they give rise are untypical of nuclei as a whole. As E. B. Wilson remarked (1925, p. 82), the commonest type of nucleus in both plants and animals is the vesicular. Now it would appear, from the important studies of Manton (1935), that the chromosomes occupy a relatively small space in a vesicular nucleus. Independent work by various authors on diverse cells suggests that as a rule the chromosomes of a vesicular nucleus are situated just below the nuclear membrane (see, for example, Ludford, 1954). When each of them has a single heterochromatic segment, it is relatively easy to make sure that these parts of the chromosomes at any rate occupy this situation (Manton, 1935). The form and position of the remainder of the chromosome cannot be defined with certainty, but it appears probable that the euchromatic segments are drawn out into threads in the same part of the nucleus. The nuclear sap occupies a large part of the volume of the nucleus, and the nucleolus (often single) is usually large.

There are all varieties of nucleus between the typical vesicular one just described, through the intermediate types investigated by Chayen and others (1953), to the kind of nucleus that results from the telophase transformation of very large chromosomes. This kind has been carefully studied by Manton (1935) in *Allium ursinum.* It appears that in this species the chromosomes maintain their early telophase positions throughout the interphase, simply swelling up to form almost the whole of the nucleus, so that there is very little room for nuclear sap. The nucleoli are not free to move, and therefore remain separate.

Pollister (1952) has very clearly described and figured two contrasting theories of nuclear structure, but it seems probable that what he describes are really the extreme forms of an object that varies widely in different plants and animals.

Of the regular constituents of the nucleus—membrane, sap, chromosomes, and nucleolus—only the chromosomes can be said to divide in ordinary mitosis. Except in certain Protozoa (see p. 474), the nuclear membrane disappears, and it then becomes impossible to distinguish sap from ground cytoplasm. It is therefore wrong to state definitely that the spindle is formed from the nuclear sap. The spindle is in fact in some cases divided by the formation of a cell-plate, or by the ingrowth of a cleavage-furrow; but there is no evidence either that it is of purely nuclear origin, or that any part of it constitutes a part of the daughter-nucleus; so that even when it is divided, it is not a continuously self-reproducing body. In the present state of knowledge it is not possible to say what parts of the cell, beyond the chromosomes, are concerned in the formation of the daughter-nuclei; but so far as is known, there is ordinarily no direct genetic relationship between the old and the new nuclear membrane, and the same applies to the nuclear sap and nucleolus.

It follows that when *Omnis nucleus e nucleo* was written in imitation of *Omnis cellula e cellula*, the similarity of the two phrases tended to hide an essential difference. The word *e* was being used in two different senses. The ground cytoplasm of a cell arises from that of a pre-existent one by mere division. Two new nuclei ordinarily arise from a pre-existent one by an entirely different process. No new nucleus will be formed unless a group of anaphase chromosomes is present, and since these chromosomes arose *e* the old nucleus, we may say that the new nucleus to that extent arose *e* the old. But the new nuclear membrane, nuclear sap, and nucleolus did not in any intelligible sense arise *e* the old.

The basic truth of the old Latin aphorisms nevertheless remains. Neither a cell nor a nucleus exists, unless there has been a pre-existing cell and nucleus which gave rise to it either directly (in the case of the cytoplasm) or indirectly (in the case of the nucleus). In Part II of this series of papers (1949) the cell was defined by its possession of protoplasm and nucleus. It is a fundamental part of the cell-theory that these are both, directly or indirectly, self-reproducing parts. The only reservations that must be made about this rule are that protoplasm must originally have evolved from matter that did not possess all its qualities, and that the nucleus, as we know it today in the great majority of plants and animals, must have evolved from a simpler structure in the distant past. The question whether it is legitimate to speak of a nucleus in the Cyanophyceae and Bacteria, and whether there is anything in those groups that throws light on the origin of the definitive nucleus, must be reserved for consideration under Proposition VI.

Although mitosis is not in fact a process of nuclear division in the great majority of plants and animals, including those in which it was first studied, yet mitotic division is a reality in certain Protozoa. Indeed, most of the errors about mitosis, entertained by the early workers on the subject, are not errors at all in the case of many flagellates. It was discovered by Blochmann (1894) that in several species of *Euglena* the nuclear membrane never disappears during mitosis, but simply becomes constricted across. Further, the nucleolus elongates into a rod thickened at each end and finally breaks in the middle, leaving one nucleolus in each of the daughter-nuclei. As Blochmann pointed out, the process is nevertheless mitotic, for chromosomes participate in it. Indeed, their behaviour is nearly normal, except that their arrangement at the stage corresponding to metaphase is less regular than usual. Keuten (1895), who had participated with Blochmann in the original work, was able to show that the chromosomes divide longitudinally and that their division-products separate in the usual way. No definite spindle or centrioles are seen. Alexeieff (1911) showed that this form of mitosis, far from being restricted to *Euglena*, occurs also in protomonads and peridinians, and indeed in certain non-flagellate Protozoa.

The more recent work of several authors, especially Hollande (1942, pp. 111–15), has confirmed the general correctness of Blochmann's and Keuten's findings. In the polymastigine, *Tetramitus*, the process is even more similar to ordinary mitosis, for a spindle is formed and the chromosomes arrange themselves very regularly at metaphase; but here again the whole affair occurs within a persistent nuclear membrane, and the nucleolus duplicates itself by division (Hollande, 1942, pp. 185–7).

Dangeard (1902) called this form of mitosis *haplomitose*, but it seems questionable whether it is simpler than ordinary mitosis. Indeed, it is doubtful whether one can make a general statement that mitosis is usually simpler in Protozoa than in other organisms, though it is certainly much more diverse. (See Grassé's admirable review of the strange process of 'pleuromitosis' in certain Protozoa (1952, pp. 104–16).) The basic facts remain that chromosomes are concerned in the formation of new nuclei, and that these chromosomes multiply by longitudinal division.

The Indirect Origin of Cells from Cells

History of the discovery

In the formulation of the cell-theory used in the present series of papers, the third proposition is this: 'Cells always arise, directly or indirectly, from pre-existing cells, usually by binary fission' (1948, p. 105). It remains to consider the indirect origin of cells from cells; that is to say, the development of a syncytium from a cell, and then of cells from the syncytium. The existence of syncytia, but not their development or transformation, has already been considered in Part III of this series (1952, pp. 177–83).

It was unfortunate that Schleiden (1838) chose a syncytium, the young endosperm, as his main subject of study when he was trying to find how cells develop. If he had chosen a tissue in which cells multiply by binary fission, it is scarcely possible that his ideas on the origin of cells would have been so erroneous. Through his influence the endosperm became a classical site for the study of the origin of cells.

Nägeli devoted much attention to cell-formation in endosperm and other syncytia. One would not suppose, from a study of his writings (1844, 1846), that binary fission was a more usual method of cellular multiplication. He performed a very useful service in demonstrating the error of Schleiden's views. The importance of his discoveries about the origin of cells in syncytia tends to be blurred for modern readers by the disproportionate emphasis he placed on his distinction between *freie* and *wandständige Zellenbildung*. In fact, however, he gave the first adequate account of the origin of cells in syncytia. He recognized that in endosperm and certain other sites there were numerous nuclei, not separated by cell-walls. (For his views on the origin of these nuclei, see above (pp. 451, 453, and 457).) The *Schleim* (protoplasm) lying between the nuclei now underwent a process of *Individualisierung* round the nuclei, and a *Membran* (cell-wall) was formed at the surface of each individualized portion of the *Schleim*. Thus, as many cells were formed as there were nuclei. Sometimes the newly-formed cells were spherical or nearly so, and free from one another and from the wall of the maternal syncytium; inevitably part of the syncytial protoplasm failed to be incorporated in the cells. This was *freie Zellenbildung*. In other cases (*wandständige Zellenbildung*) no protoplasm was left out of the new cells, for the cell-walls were formed in immediate apposition to one another (or, in the external part of the syncytium, to the maternal cell-wall).

Rathke and Kölliker were the first to describe the origin of cells from syncytia in animals. I have already published a translation of Rathke's words (1952, p. 180). He remarks very tersely that in the embryo of Crustacea, nuclei are formed for the future embryonic cells before the cells themselves originate. He does not mention the mode of formation of the nuclei.

In his study of the development of cephalopods, Kölliker (1844) realized that the *Furchungssegmente* (*blastocones* of Vialleton, 1888) were not separated from the yolk by any distinct boundary. He must have realized, therefore, that the *Embryonalzell* (nucleus) of one blastocone was not separated from that of the next by any membrane; or, to put it in other words, he must have understood that he was dealing with a syncytium. He knew that the nuclei duplicated themselves in the blastocones, and that uninucleate cells (the *Furchungskugeln* or blastomeres) were budded off from their tips. It follows that he described the origin of cells from a syncytium, though this fact is obscured by his strange nomenclature, which I have already explained (1953, p. 418).

Incomplete cleavage is usually called meroblastic, but Remak himself (1852), when he introduced the term *méroblastique*, did not attach exactly this significance to it. For him, an egg was meroblastic if the embryo was formed from only a part of it: if the whole egg clove to convert itself into the embryo, it was holoblastic. It seems uncertain whether the egg of cephalopods would be meroblastic by Remak's definition, despite the fact that the cleavage-furrows do not pass right through it. The meaning of the term that is usual today appears to derive from Nicholson (1870, p. 217), who wrote of the development of the lobster, 'The ovum is "meroblastic", a portion only of the vitellus undergoing segmentation.'

The first really adequate description of the origin of cells from a syncytium in animals was given by Leuckart (1858, pp. 210–11), in his account of the development of *Melophagus* (Diptera Pupipara). He tentatively derived the nuclei of the early embryo from the germinal vesicle of the egg. He remarks that the development of the egg of insects conforms to the usual process of embryonic cell-formation, but 'A difference appears to me to exist here, only in so far as in insects the envelopment of the cell-nuclei with yolk-substance first occurs late, after the number of nuclei has already been considerably increased, while in other cases such an envelopment happens from the beginning, so that the division of the nuclei has then for a consequence, naturally and also constantly, a division of the yolk'. Robin (1862) noticed this method of cell-formation in various culicines; he called it *gemmation* and distinguished it from cleavage. He did not remark, however, on the presence of one nucleus in each of the cells formed by this process. Weismann confirmed Leuckart's findings by his studies of the development of *Chironomus* (1863*a*, pp. 112–13) and *Musca* (1863*b*, p. 163). He noticed the rising of the nuclei to the *Keimhautblastem* (blastoderm) in *Musca*, and compared it to the rising of air-bubbles to the surface of water. When the blastoderm had separated itself off round each of these nuclei, the newly-formed cells multiplied by ordinary binary fission.

Comment

Syncytia that eventually resolve themselves into cells do not constitute an exception to the cell-theory as formulated in this series of papers. Particular parts of the body are often permanently syncytial. Not many groups of organisms are wholly syncytial, even in their somatic tissues. The belief that rotifers provide an example will not withstand critical examination, though many of their organs are wholly or partly syncytial (Martini, 1912; Nachtwey, 1925). The same applies to nematodes.

In the great groups of syncytial plants, the Siphonales, Cladophorales, and Phycomycetes, there is nearly always a periodical reversion to the haplocyte or diplocyte; that is to say, to the *cell* as defined in this series of papers (Part III, 1952). It will be recalled that the gametes of the Siphonales are generally flagellate cells; in *Vaucheria*, in which they are not flagellate, they are nevertheless cells. In the Cladophorales, asexual reproduction is in nearly every case by zoospores in the form of flagellate cells.

The two groups of Phycomycetes differ markedly in their reproductive processes. In the Oomycetes, asexual reproduction is generally by flagellate cells, sexual reproduction by the fusion of uninucleate protoplasmic masses from the antheridium with uninucleate ova, or, in the more primitive forms (Uniflagellatae), generally by the fusion of uninucleate flagellate gametes. In certain Zygomycetes, however, the cellular phase seems genuinely to have been lost. In *Pilobus crystallinus* the sporangiospores are in fact uninucleate, but in some other species each spore is multinucleate, so that asexual reproduction occurs without the intervention of a cellular phase. This applies, for instance, to *Rhizopus nigricans* and *Sporidinia grandis*. Now it is characteristic of sexual reproduction in Zygomycetes that the whole of the syncytial protoplasm of one gametangium fuses with that of another, with subsequent fusion of the nuclei in pairs. The new individual produced by this fusion proceeds to asexual reproduction (without any intercalated cellular phase) by forming multinucleate sporangiospores. The cycle is thus completed without the existence of a cell at any stage of the life-history.

Such forms as *Rhizopus* and *Sporodinia* are of exceptional interest to the student of the cell-theory. Their existence is a disproof that the theory is of universal application. It is to be remembered, however, that we can quote few similar examples in plants, and nothing at all that is even remotely similar in animals, except in certain Ciliophora. A discussion of the latter is reserved for a future paper in this series, which will be devoted to a consideration of the status of the Protozoa from the point of view of the cell-theory. (I have already treated the subject shortly (1948*a*).)

How has it come about in the course of evolution that the great majority of organisms consist largely of cells or at least are derived from and return to cells in the course of their life-histories? This is one of the fundamental questions of biology, yet there is not very much that can be said in answer to it.

It is necessary to consider the somatic tissues separately from the gametes, because quite different causes appear to have been at work. In the case of the somatic tissues it is to be noted that a high degree of complexity of structure is never reached in a wholly syncytial organism. The repeatedly-quoted case of *Caulerpa* (Siphonales) is misleading, for the parts that superficially resemble the leaf, stem, and root of higher forms do not attain to anything approaching the degree of differentiation that their external aspect would suggest. It seems that organisms can more easily achieve differentiation in a cellular tissue than in a syncytial mass of protoplasm. This may perhaps be correlated with the obvious fact that synthesized substances are more easily localized in cellular tissues. It might perhaps be possible to devise experiments to discover something about the advantages an organism obtains by keeping its protoplasm in amounts small enough for each to be related to a single nucleus.

The reason why even a somatically syncytial organism nearly always has unicellular gametes is of quite a different nature. Why should not a higher animal, for instance, reproduce by syncytial gametes, like those of *Rhizopus*? Let us suppose that the nuclei of the syncytial gametes of such an animal were the immediate products of meiosis. They would necessarily differ among themselves in their gene-complexes. When karyogamy had occurred, a wide assortment of gene-complexes would be present in the embryo. Let us suppose, for example, that one of these complexes was such as to be potentially favourable to the survival of the organism, if present in nerve-cells. It will at once be evident that the nuclei derived from the zygote nucleus carrying that particular complex might be absent from the nervous system and present only in other tissues, in which it could not exhibit its beneficial effects. There would only be two ways of overcoming this barrier to the action of natural selection and therefore to evolution. One possibility would be to form a large number of uninucleate spores, each capable of developing into a whole organism carrying the same gene-complex in every nucleus of the somatic tissues. (This is exactly what some of the Zygomycetes do—and without wasting much time on vegetative growth at this stage.) A much simpler and quicker way would be to reproduce sexually by uninucleate gametes.

Natural selection can only act effectively on an organism that has the same gene-complex in the nuclei of all its somatic tissues; and that can only be achieved by periodical reversion to the unicellular state.

I thank Professor A. C. Hardy, F.R.S., for his continued support and encouragement of cytological studies. Dr. C. F. A. Pantin, F.R.S., has kindly given me the benefit of his advice. I take the opportunity of acknowledging Mrs. J. A. Spokes's secretarial help with this series of papers. The Royal Society, The Linnean Society, and the Department of Botany, Oxford, have kindly allowed copies to be made of illustrations in books and journals in their possession. Mr. Michael Lyster has made most of the photographs used in this paper. Several of them are clearer than the originals, from increase of contrast, but none has been retouched.

REFERENCES

ALEXEIEFF, A., 1911. Compt. rend. Soc. Biol., **71**, 614.
AUERBACH, L., 1876. Centralbl. med. Wiss., **14**, 1.
BAER, — v., 1846. Bull. Classe phys.-math. Acad. Imp. St. Pétersbourg, **5**, col. 231.
BAGGE, H., 1841. *Dissertatio inauguralis de evolutione Strongyli auricularis et Ascaridis acuminatae viviparorum.* Erlangae (ex officina Barfusiana).
BAKER, J. R., 1948*a*. Nature, **161**, 548 and 857.
—— 1948*b*. Quart. J. micr. Sci., **89**, 103.
—— 1949. Ibid., **90**, 87.
—— 1952. Ibid., **93**, 157.
—— 1953. Ibid., **94**, 407.
BALBIANI, —, 1876. Compt. rend. Acad. Sci., **83**, 831.
BARRY, M., 1839. Phil. Trans., **129**, 307.
—— 1841*a*. Ibid., **131**, 193.
—— 1841*b*. Ibid., **131**, 201.
—— 1841*c*. Ibid., **131**, 217.
BENEDEN, É. VAN, 1875. Bull. Acad. roy. Belg., **40**, 686.
—— 1876. Ibid., **42**, 35.
BLOCHMANN, F., 1894. Biol. Centralbl., **14**, 194.
BOVERI, T., 1888. *Zellen-Studien. Heft 2. Die Befruchtung und Teilung des Eies von Ascaris megalocephala.* Jena (Fischer).
—— 1904. *Ergebnisse über die Konstitution der chromatischen Substanz des Zellkerns.* Jena (Fischer).
BREUER, R., n.d. (1844). *Meletemata circa evolutionem ac formas cicatricum.* Editio altera. Vratislaviæ (Schumann).
BÜTSCHLI, O., 1873. Arch. mikr. Anat., **9**, 657.
—— 1875. Zeit. wiss. Zool., **25**, 201.
—— 1876. Abh. Senckenb. naturf. Ges. (Frankfurt a. M.), **10**, 213.
CHAYEN, J., DAVIES, H. G., and MILES, J. J., 1953. Proc. Roy. Soc. B, **141**, 190.
DANGEARD, P. A., 1902. Botaniste (no vol. number), 97.
EBERTH, C. J., 1876. Arch. path. Anat. Physiol. klin. Med. (Virchow), **67**, 523.
EHRENBERG, D. C. G., 1838. *Die Infusionsthierchen als vollkommene Organismen.* Leipsig (Voss).
EWETSKY, T. v., 1875. Untersuch. path. Inst. Zürich, **3**, 89.
FLEMMING, W., 1877. Arch. mikr. Anat., **13**, 693.
—— 1878. Schrift. naturwiss. Ver. Schleswig-Holstein, **3**, 23.
—— 1879*a*. Arch. path. Anat. Physiol. klin. Med. (Virchow), **77**, 1.
—— 1879*b*. Arch. mikr. Anat., **16**, 302.
—— 1880. Ibid., **18**, 151.
—— 1882*a*. Ibid., **20**, 1.
—— 1882*b*. *Zellsubstanz, Kern und Zelltheilung.* Leipsig (Vogel).
GEGENBAUR, C., 1858. Abh. naturf. Ges. Halle, **4**, 1.
GOODSIR, J. and H. D. S., 1845. *Anatomical and pathological observations.* Edinburgh (MacPhail).
GRASSÉ, P.-P. (edited by), 1952. *Traité de Zoologie. Anatomie, Systématique, Biologie.* Paris (Masson).
GÜNSBURG, F., 1848. *Studien zur speciellen Pathologie.* Zweiter Band. Leipsig (Brockhaus).
HAECKEL, E., 1866. *Generelle Morphologie der Organismen.* Vol. 1. Berlin (Reimer).
HANSTEIN, —, 1870. Sitz. niederrhein. Ges. Bonn (no vol. number), 217.
HEIDENHAIN, M., 1894. Arch. mikr. Anat., **43**, 423.
HENLE, J., 1841. *Allgemeine Anatomie. Lehre von den Mischungs- und Formbestandtheilen des menschlichen Körpers.* Leipsig (Voss).
—— 1866. *Handbuch der Eingeweidelehre des Menschen.* Braunschweig (Vieweg).

HERTWIG, O., 1876. Morph. Jahrb., **1**, 347.
—— 1877. Ibid., **3**, 1.
—— 1878*a*. Ibid., **4**, 156.
—— 1878*b*. Ibid., **4**, 177.
HOFMEISTER, W., 1848. Bot. Zeit., **6**, col. 425.
—— 1848*b*. Ibid., **6**, col. 649.
—— 1867. *Die Lehre von der Pflanzenzelle.* Leipsig (Engelmann).
HOLLANDE, A., 1942. Arch. Zool. exp. gén., **83** (2), 1.
KEUTEN, J., 1895. Zeit. wiss. Zool., **60**, 214.
KÖLLIKER, A., 1843. Arch. Anat. Physiol. wiss. Med. (no vol. number), 68.
—— 1844. *Entwickelungsgeschichte der Cephalopoden.* Zürich (Meyer & Zeller).
KOWALEVSKI, A., 1871. Mém. Acad. imp. Sci. St. Pétersbourg, **16**, No. 12.
KRAUSE, W., 1870. Centralbl. med. Wiss., **8**, 383.
—— 1870. Nachr. Kön. Ges. Wiss. Univ. Göttingen (no vol. number), 140.
—— 1876. *Handbuch der menschlichen Anatomie.* 3rd ed., vol. 1. Hannover (Hahn).
KROHN, A., 1852. Arch. Anat. Physiol. wiss. Med. (no vol. number), 312.
LÁNG, E., 1872. Arch. path. Anat. Physiol. wiss. Med. (Virchow), **54**, 85.
LANKESTER, E. R., 1875. Quart. J. micr. Sci., **15**, 37.
LEUCKART, R, 1858. Abh. naturf. Ges. Halle, **4**, 145.
LUDFORD, R. J., 1954. Brit. Journ. Cancer, **8**, 112.
LUNDEGARDH, H., 1913. Arch. Zellforsch., 9, 205.
MANTON, I., 1935. Proc. Roy. Soc. B, **118**, 522.
—— 1950. *Problems of cytology and evolution in the Pteridophyta.* Cambridge (University Press).
MARTINI, E., 1912. Zeit. wiss. Zool., **102**, 425.
MAYZEL, W., 1875. Centralbl. wiss. Med. (no vol. number), 849.
—— 1877. Ibid., **15**, 196.
NACHTWEY, R., 1925. Zeit. wiss. Zool., **126**, 239.
NÄGELI, C., 1841. Verh. schweiz. naturf. Ges. Zürich, **26**, 85.
—— 1842. *Zur Entwickelungsgeschichte des Pollens bei den Phanerogamen.* Zürich (Ovell & Füssli).
—— 1844. Zeit. wiss. Bot., **1**, 34.
—— 1844. Ibid., **1** (1), 34.
—— 1846. Ibid., **1** (3–4), 22.
NICHOLSON, H. A., 1870. *A manual of zoology for the use of students.* London (Hardwicke).
OELLACHER, J., 1872. Zeit. wiss. Zool., **22**, 373.
PEREMESCHKO, —, 1878. Centralbl. med. Wiss., **16**, 547.
POLLISTER, A. W., 1952. Exp. Cell Res., Suppl. **2**, 59.
RABL, C., 1885. Morph. Jahrb., **10**, 214.
RATHKE, H., 1844. *De animalium crustaceorum generatione.* Regiomonti (Dalkowski).
REICHERT, K. B., 1846. Arch. Anat. Physiol. wiss. Med. (no vol. number), 196.
—— 1847. Ibid. (no vol. number), 88.
REMAK, R., 1841. Med. Zeit., **10**, 127.
—— 1845. Neue Not. Geb. Natur- und Heilk. (Froriep), **35**, col. 305.
—— 1852. Compt. rend. Acad. Sci., **35**, 341.
—— 1855. *Untersuchungen über die Entwickelung der Wirbelthiere.* Berlin (Reimer).
—— 1862. Arch. Anat. Physiol. wiss. Med. (no vol. number), 230
ROBIN, C., 1862. Compt. rend. Acad. Sci., **54**, 150.
RUSSOW, E., 1872. Mém. Acad. Imp. Sci. St. Pétersbourg, **19**, 1.
SCHLEICHER, W., 1878. Centralbl. med. Wiss., **16**, 418.
—— 1879. Arch. mikr. Anat., **16**, 248.
SCHLEIDEN, M. J., 1838. Arch. Anat. Physiol. wiss. Med. (no vol. number), 137.
SCHNEIDER, A., 1873. Ber. Oberhess. Ges. Natur-und Heilk., **14**, 69.
SCHULTZE, M., 1861. Arch. Anat. Physiol. wiss. Med. (no vol. number), 1.
SCHWANN, T., 1838. Neue Not. Geb. Natur- und Heilk. (Froriep), **5**, col. 33.
—— 1839. *Mikroskopische Untersuchungen.* Berlin (Sander'schen Buchhandlung).
SPENGEL, J. W., 1876. Arb. Zool.-zoot. Inst. Würz., **3**, 1.
STRASBURGER, E., 1875. *Uerer* [*sic*] *Zellbildung und Zelltheilung.* 1st ed. Jena (Dabis).
—— 1876. Ibid. 2nd ed. Jena (Dabis).
—— 1877. Jen. Zeit. Naturwiss., **11**, 435.

STRASBURGER, E., 1879. Sitz. Jena Ges. Med. Naturw. (no vol. number), 93.
—— 1880. *Zellbildung und Zelltheilung*. 3rd ed. Jena (Fischer).
—— 1884. Arch. mikr. Anat., **23**, 246.
TSCHISTIAKOFF, J., 1875*a*. Bot. Zeit., **33**, cols. 1 and 17.
—— 1875*b*. Ibid., **33**, col. 80.
VALENTIN, G., 1835. *Handbuch der Entwickelungsgeschichte des Menschen*. Berlin (Rücker).
—— 1842. Article on *Gewebe des menschlichen und thierischen Körpers* in WAGNER, 1842, p. 617.
VIALLETON, L., 1888. Ann. Sci. nat. Zool., **6**, 165.
VIRCHOW, R., 1857. Arch. path. Anat. Physiol. klin. Med. (Virchow), **11**, 89.
WAGNER, R., 1842. *Handwörterbuch der Physiologie mit Rücksicht auf physiologische Pathologie*. Vol. 1. Braunschweig (Vieweg).
WALDEYER, W., 1888. Arch. mikr. Anat., **31**, 1.
WEISMANN, A., 1863*a*. Zeit. wiss. Zool., **13**, 107.
—— 1863*b*. Ibid., **13**, 159.
WILSON, E. B., 1925. *The cell in development and heredity*. New York (Macmillan).

DATE DUE

DEC 06 2004	